A Guide to Britain's Rarest Plants

Christopher J. Dixon

Pelagic Publishing
www.pelagicpublishing.com

Published by Pelagic Publishing
www.pelagicpublishing.com
PO Box 725, Exeter, EX1 9QU, UK

A Guide to Britain's Rarest Plants

ISBN 978-1-78427-146-6 (Pbk)
ISBN 978-1-78427-147-3 (ePub)
ISBN 978-1-78427-148-0 (Mobi)
ISBN 978-1-78427-149-7 (PDF)

British Library Cataloguing in Publication Data
A catalogue record for this book is available from the British Library

Cover photograph: Perennial centaury, *Centaurium scilloides*

Printed and bound in India by Replika Press Pvt. Ltd.

Contents

Introduction

The flora of the British Isles is exceptionally well studied, largely through the efforts of a small army of enthusiastic amateurs. By recording every species that occurs in each cell of a grid drawn across Great Britain, Ireland and adjacent islands, comprehensive maps have been compiled and revised, showing the range of each species. The most-used grid is at the scale of 10 km × 10 km squares, but similar work is being carried out at the scale of 2 km × 2 km grid squares, and even 1 km × 1 km squares. For the vast majority of species, their current and past distributions are therefore known to a very high degree of accuracy, enabling detailed assessments to be made of any changes in distribution due to changes in land use or climate.

At the same time, the urban population has continued to increase, and knowledge of the natural world has continued to decline. A 2013 survey by the Woodland Trust found that 43% of adults could not recognise an oak leaf, and only around one in six could recognise an ash leaf. If that is the case for two of our most abundant and distinctive trees, then knowledge of smaller and rarer species must be far scanter. Despite the need for biologists to know which species is which in order to study them at all, natural history is hardly taught in universities any more. Knowledge of natural history is rapidly fading among both professionals and the general public.

On a global scale, Britain is no biodiversity hotspot – it doesn't have the exceptional diversity of a tropical rainforest or the South African *fynbos* – but it is still unquestionably diverse. There is a huge range of habitats, from salt-marshes to meadows, woodlands to high-mountain crags. Of the 3,000 or so plant species in the British Isles, many are well known, but there are others that you may never have heard of.

While some plants grow almost everywhere across the region, others are found at only a small handful of sites. Plants with restricted distributions can be a source of local pride. Several of the species in this book were chosen to represent their counties in Plantlife's *County Flowers* programme, for instance. Rare plants such as these are often held to be particularly special. Unless you go specifically looking for them, you are very unlikely to come across one. Many of our rarest plants are hard to spot in the first place, or are difficult to distinguish from similar species once they've been found. For most of them, Britain is only part of their overall range, but in a few cases, this is all they've got. In every case, there is a story to tell. (How did the species arrive in Britain? Why has it dwindled to so few sites? How can we ensure it survives?) In telling the story of Britain's rarest plants, we are telling the story of Britain's natural world, the turmoil it has faced in the past, and the challenges it will meet in the future.

COVERAGE

The first challenge in presenting a work like this is to decide which plants to include. Any such list is to some degree arbitrary, and this book is no exception. The species here are limited to vascular plants (ferns, conifers and flowering plants, including grasses and sedges, but no mosses or seaweeds) that are generally considered to be native to Scotland, England or Wales. It is not always easy to tell which species are native (i.e. naturally occurring) and which are introduced (only present through human actions), and some instances where there is a significant possibility of human interference have been excluded. (In general, native species are expected to have a fairly stable distribution, to occur in natural habitats, especially those far from habitation, and so on; introduced species are more likely to suddenly expand or contract their ranges, and to occur predominantly in artificial habitats close to human habitation.) These include the lesser tongue-orchid (*Serapias parviflora*), which grew on a Cornish cliff-top for several years, and the interrupted brome (*Bromus interruptus*), which has become extinct in the wild.

Population estimates are hard to produce accurately, particularly given that plants can spread vegetatively, making it unclear how many separate plants are present in a population. It is therefore more reliable to use a proxy measure of rarity. This book covers a total of 66 species that are native to Britain and each of which is found in no more than three 10 km × 10 km grid squares out of the nearly 3,000 that cover England, Wales and Scotland.

Rare subspecies of more widespread species have also been left out; these include the English sandwort (*Arenaria norvegica* subsp. *anglica*), the South Stack fleawort (*Tephroseris integrifolia* subsp. *maritima*) and our endemic subspecies of the common mouse-ear (*Cerastium fontanum* subsp. *scoticum*), to name just a few examples. In each of these cases, another subspecies of the same species is more widespread, meaning that the species as a whole is not rare enough to be included.

Some plants undergo a process known as apomixis, whereby offspring are almost always produced asexually. The usual definition of a species doesn't really apply in these cases, and their reproductively isolated groups – known as microspecies or agamospecies – are large in number but generally hard to distinguish from their relatives. In order to assign a particular plant to a microspecies, it is often necessary to consider the variation within the wider population, because there is so much overlap between similar microspecies. These 'critical groups', containing numerous barely distinguishable taxa, include the brambles (*Rubus* spp.), lady's-mantles (*Alchemilla* spp.), whitebeams (*Sorbus* spp.), hawkweeds (*Hieracium* spp.), sea-lavenders (*Limonium* spp.), eyebrights (*Euphrasia* spp.) and dandelions (*Taraxacum* spp.). I have not described the many rare microspecies in detail in the main part of the book, but there is a short discussion of each of the critical groups towards the end.

IDENTIFICATION

I have tried to include some indication of how each of the species differs from its nearest relatives in the area. These descriptions are necessarily sketchy, and can never be enough for formal identification. Anyone who wants to identify the plants they find with greater certainty should invest in a good field guide; Stace's *New Flora of the British Isles* (3rd edition, 2010, Cambridge University Press) is the standard text in this area and is highly recommended.

For each of the species examined, I have also given the typical flowering period and conservation status. The timing of flowering of any plant depends on a number of factors, including temperature and day length, many of which vary from year to year. Some individuals are likely to flower outside the stated period even in a typical year, and in atypical years, the entire population may flower earlier or later than expected. The conservation status uses the categories created by the International Union for Conservation of Nature (IUCN), and refer here only to the species' occurrence within the United Kingdom (LC, least concern; NT, near threatened; VU, vulnerable; EN, endangered; CR, critically endangered; EX, extinct). Many of our rare species are more endangered here than in other parts of their range, and it is inevitable that a species is at greater risk of being lost from Great Britain than of going extinct globally. These conservation assessments are taken from the *Conservation Designations for UK Taxa* published by the Joint Nature Conservation Committee (JNCC). Below each species account, there is a short section labelled 'Also in the area'; these give examples of some other rare plants that grow nearby, where applicable.

LEGAL ISSUES AND CONSERVATION

Many of the species described in this book are heavily protected under the Wildlife and Countryside Act 1981. Plants listed on Schedule 8 of the act may not be uprooted, picked or destroyed, and no material from these plants, whether alive or dead, can be sold or even advertised for sale. This is merely the legal situation, however. Conservation of these vulnerable species requires us to treat them very carefully indeed. Even those not directly protected by law should never be picked or damaged. If you are lucky enough to find one of these species in the wild, by all means take photographs, but even then, be careful where you tread for fear of damaging individuals you may not have noticed. Two of the species are also listed under Appendix II of the EU Habitats Directive, granting them protection throughout the European Union: creeping marshwort, *Apium repens* (p. 6); and lady's-slipper orchid, *Cypripedium calceolus* (p. 42).

There are examples in this book of species that have been saved from extinction through the collection of seeds and being grown in cultivation. Although well-intentioned, this is not something to be generally encouraged. Seed set, survival and germination are the riskiest parts of a plant's life cycle, and reducing the number of seeds available to produce the next generation is very likely to harm the population. If you would like to get involved in the conservation efforts around these and other species, the best approach is to contact your local wildlife trust (http://www.wildlifetrusts.org/) or the relevant statutory agencies: Natural England, Natural Resources Wales, and Scottish Natural Heritage.

IMAGES

Except where otherwise noted, all images are copyright of the author ('CJD'). Most of those from other sources are used under the terms of a Creative Commons Attribution Share-Alike licence; they can be freely reproduced and modified, provided the same licence is applied to any derivatives. Their sources are indicated in each case:

EM: eMonocot, http://www.e-monocot.org/

GG: Geograph Britain and Ireland, http://www.geograph.org.uk/

IC: Il Cercapiante, http://dbiodbs.units.it/carso/cercapiante01

TB: Tela Botanica, http://www.tela-botanica.org/

WC: Wikimedia Commons, http://commons.wikimedia.org/

Full URLs are given on p. 143. The few remaining images are reproduced with the kind permission of their authors.

Creeping marshwort

Apium repens

Creeping marshwort is a low-growing herb in the same family as carrots, parsnips, parsley and celery, with tiny white flowers less than 2 mm across. It is closely related to celery, which grows wild in many coastal parts of England and Wales. It is even more closely related to fool's water-cress, *Apium nodiflorum*, which is usually a medium-sized plant, but can adopt a low-growing habit when it is heavily grazed, and can occur alongside creeping marshwort. The two can even interbreed to produce a hybrid. The main flower-stalk of creeping marshwort is longer than the secondary stalks, and the segments of each compound leaf are not noticeably longer than they are wide, both of which separate creeping marshwort even from short forms of fool's water-cress.

Creeping marshwort is widespread, but not common, in Central and Southern Europe. In Britain, it is restricted to a few sites on the flood plains of the River Thames around Oxford, chiefly the ancient Port Meadow, where it was first found by the botanist John Sibthorp in 1794. These sites are often covered with water for long periods over the winter, and are heavily grazed by cattle and horses. As well as reproducing via seeds, creeping marshwort can spread vegetatively, with pieces of the plant floating on the floodwaters to find new habitat downstream.

A population of creeping marshwort has been introduced further downstream, although a proposed new channel that would reduce the risk of flooding in Oxford may require that population to be re-transplanted.

ALSO IN THE AREA

At the northern end of Port Meadow lie the ruins of Godstow Abbey, a former Benedictine nunnery. The nuns' herb-garden included birthwort, *Aristolochia clematitis*, which survives in the grounds of the abbey.

CONSERVATION STATUS	VU
FLOWERING	Jul–Aug
PLANT HEIGHT	2–10 cm

Creeping marshwort in flower. (CJD)

Horses grazing on an inundated Port Meadow in winter. (CJD)

Alpine rock-cress

Arabis alpina

Alpine rock-cress is a mat-forming perennial plant that produces flowering stalks up to 40 cm (16 in) tall. It is classified in the same family as mustard, cress and cabbages - Brassicaceae. This family also includes a number of similar plants, and Alpine rock-cress can be told apart from its relatives only by a combination of several factors.

Alpine rock-cress has an arctic-alpine distribution, meaning that it is found at high latitudes in Scandinavia and elsewhere, and also in the similar climate of high mountains further south, including the Alps and the Pyrenees, but not in the areas in between. In Britain, it sometimes appears as a garden escape, but occurs natively only on the Isle of Skye.

Alpine rock-cress was discovered on Skye by Henry Chichester Hart, a colourful Irishman of independent means. Hart was an indefatigable field botanist, and happy to explore areas that were too difficult for almost anyone else. (He once bet a friend that he could walk the 69 miles - 111 km - from a tram stop in Dublin to the highest point of the Wicklow Mountains and back in a day; he won the bet.) This may be why he was the first to come across Alpine rock-cress, given the steepness of the Black Cuillin range, where it grows at altitudes of 820–850 m (2,700–2,800 ft). The plants' precise locations are kept secret to save the population from harm. The extreme conditions of the Cuillin also hinder research into Alpine rock-cress, including tasks as simple as surveying the population size. Plants cultivated from the native population were re-introduced in 1975 in an attempt to bolster the population. In 1993, only three colonies were found, totalling 83 plants, and it is not known how the species has fared since.

ALSO IN THE AREA

Another Arctic plant, Iceland purslane (*Koenigia islandica*) from the dock family, occurs on the Trotternish peninsula of Skye as well as in some locations on Mull. It is very inconspicuous, and was not noticed in Britain until 1934.

CONSERVATION STATUS	EN
FLOWERING	Jul–Aug
PLANT HEIGHT	5–40 cm

Alpine rock-cress growing in the Apennines. (CJD)

Coire Lagan in the Black Cuillin. (Russell Wills; GG)

Bristol rock-cress

Arabis scabra

Bristol rock-cress is closely related to Alpine rock-cress, *Arabis alpina* (p. 8), and is similar in appearance, although it forms compact tufts up to 25 cm (10 in) high instead of sprawling mats, and has shiny leaves with a sparse covering of thick bristles. It grows on open parts of rocky limestone slopes, and is readily smothered by competition. Each stem produces up to a dozen flowers, each 5–6 mm in diameter.

In the wild, Bristol rock-cress grows in the mountains of northern Spain and southern France – just extending over the border into Switzerland near Geneva – and at one site in Britain. That site is the Avon Gorge, where it was first found in 1686 by James Newton, assistant to the parson-naturalist John Ray, on St. Vincent's Rock – a site now better known as the foundation for the Clifton Suspension Bridge. Bristol rock-cress grows chiefly on south-facing cliffs on either side of the River Avon, with smaller populations in quarries further downstream at Shirehampton. The populations by the bridge may have been briefly harmed by toxic heavy metals that were present in the copper slag used to shot-blast the bridge in 1995. A greater threat to the plant's continued survival may come indirectly from the road at the bottom of the gorge. In order to prevent rock-falls onto the road, strong nets have been strung across some of the slopes, trapping leaf-litter and allowing more vigorous plants to grow and outcompete the rather retiring Bristol rock-cress.

Bristol rock-cress has been introduced to similar sites nearby, but few of the populations have survived. One exception is a population at Combwich on the tidal reaches of the River Parrett in Somerset.

ALSO IN THE AREA

The Avon Gorge contains a number of rare species, including honewort (*Trinia glauca*) and several endemic or near-endemic whitebeam microspecies (*Sorbus leighensis*, *S. eminentiformis*, *S. whiteana*, *S. wilmottiana* and *S. bristoliensis*).

CONSERVATION STATUS	VU
FLOWERING	Apr–May
PLANT HEIGHT	10–25 cm

Bristol rock-cress growing in France. (Joceline Chappert-Bessière; TB)

Cliffs of the Avon Gorge. (CJD)

Norwegian mugwort

Artemisia norvegica

Norwegian mugwort is a small herb related to other mugworts and wormwoods, in the tremendously diverse daisy or sunflower family. The family relationship becomes clear when you examine the tiny drooping flower-heads, which resemble miniature sunflowers. It differs from other *Artemisia* species in its tiny size – only reaching 8 cm (c. 3 in) tall – and by its small number of flower-heads, typically only 1 or 2 per stem.

Norwegian mugwort grows in the northern Ural Mountains of Russia, in Norway, and in western Scotland. A closely related species (sometimes considered part of the same species) occurs in western North America and the Russian Far East. The Scottish populations are restricted to three mountains around Ullapool, where they grow along the high ridges above 700 m (2,300 ft). The first of these was found by the historian Sir Christopher Cox in 1950, the year he was knighted for services to education. Despite being such a recent discovery, the Scottish plants are unlikely to be a recent arrival, as they differ consistently from plants in the Urals and Norway by their shorter stature and less divided leaves. They have even been recognised as a separate variety, *Artemisia norvegica* var. *scotica*. Even such a seemingly small amount of evolution is likely to have taken hundreds or thousands of years. The fact that the Scottish populations grow so high up on the mountains also means that they will struggle to cope with climatic warming, because there is nowhere higher that they could migrate to in order to reach colder conditions. Indeed, they may already be struggling; it is reported that the plants rarely produce much viable seed.

ALSO IN THE AREA

Dactylorhiza incarnata subsp. *cruenta*, a subspecies of the early marsh orchid (sometimes treated as a separate species, *Dactylorhiza cruenta*) grows in the mires east of Lochinver.

CONSERVATION STATUS	VU
FLOWERING	Jul–Sep
PLANT HEIGHT	3–8 cm

Norwegian mugwort. (CJD)

Bare ground with eroding sandstone on Cùl Mòr. (CJD)

Pedunculate sea-purslane

Atriplex pedunculata

Pedunculate sea-purslane is a shrubby, mealy plant that grows up to 30 cm (12 in) tall. It is classified in the family Amaranthaceae (formerly Chenopodiaceae), and is related to the much more widespread sea-purslane (*Atriplex portulacoides*), and more distantly to the oraches, goosefoots, spinach and beet. It can be immediately told apart from all its British relatives by its long-stalked fruits with two conspicuous bulges on the sides, a bit like the fruits of shepherd's purse, *Capsella bursa-pastoris*.

Worldwide, pedunculate sea-purslane is found in coastal areas from the Gulf of Bothnia to France, as well as around the Black Sea and at saline sites inland. In the past, it has occurred at sites from Lincolnshire to Kent, but was not found anywhere in Britain from 1938 to 1987, when a new population was discovered in southern Essex. Pedunculate sea-purslane is most abundant around the coasts of Denmark, and it has been speculated that it was unwittingly re-introduced to Britain by migratory brent geese.

The abundance of pedunculate sea-purslane is known to vary wildly from one year to the next, making local extinction a very real risk if few seeds are produced one year, or fail to settle in suitable positions. To improve its chances of persisting in the longer term, pedunculate sea-purslane has therefore been introduced to other sites in the area, including some on military land, where it can be protected more easily. It could also reappear at any time at any suitable sites along England's east coast that are visited by migrating geese.

ALSO IN THE AREA

Saltmarsh goosefoot, *Chenopodium chenopodioides*, is largely restricted in Britain to the drier parts of salt-marshes the Thames Estuary. See also least lettuce (*Lactuca saligna*, p. 72).

CONSERVATION STATUS	CR
FLOWERING	Aug–Sep
PLANT HEIGHT	10–30 cm

Growing in Picardy, France. (Hugues Tinguy; TB)

Salt-marsh in Essex. (CJD)

Small hare's-ear

Bupleurum baldense

Small hare's-ear is a member of the carrot and parsnip family, but differs from most members of that family in having simple, rather than deeply divided, leaves. Because of that, and its tiny yellow flowers surrounded by yellow-green bracts, it can look rather like the dwarf spurge, *Euphorbia exigua*. It can grow up to 25 cm (10 in) tall but is typically only a few centimetres tall, and flowers from May to July; fruiting finishes by September.

Several other species in the same genus occur sporadically in Britain, but among them only the slender hare's-ear, *Bupleurum tenuissimum*, is native here. It differs from the small hare's-ear in having narrower bracts that cannot conceal the flowers, and warty fruits.

Small hare's-ear was first found in Britain around 1801 by the Rev. Aaron Neck near Torquay, where it has since died out. It now grows in coastal grassland grazed by rabbits at two sites in southern England. One, containing up to a few hundred plants, is at Beachy Head in Sussex; the other is over 150 miles to the west near Brixham in Devon, and may boast 2,000 individuals in a good year. These are the species' northernmost outposts: on the continent it occurs across Italy, Spain, France and the Channel Islands. Small hare's-ear generally grows over calcareous rocks, and on the continent it is not limited to coastal locations.

The greatest threat to the small hare's-ear in Devon may be from people trampling the plants underfoot. In Sussex, the crumbling cliffs recede by 25 cm (10 in) a year and remove a large part of the population of small hare's-ears in the process, leaving only seeds that were fortuitously blown a short distance inland to produce the next year's plants.

ALSO IN THE AREA

The eastern end of the South Downs houses the only British localities of the spiked rampion, *Phyteuma spicatum*.

CONSERVATION STATUS	VU
FLOWERING	Jun–Jul
PLANT HEIGHT	2–10 cm

Small hare's-ear at Beachy Head. (CJD)

Crumbling cliff edge at Beachy Head. (CJD)

Club sedge

Carex buxbaumii

Club sedge is one of the more than 70 species of sedge (genus *Carex*) growing in the British Isles. Sedges are a difficult and much overlooked group, and the various species can be difficult to tell apart, even for experienced botanists. At first glance, club sedge resembles the common sedge, *Carex nigra*, but its flowering spikes are structured quite differently: they are all of similar appearance, with male flowers at the base of each spike, and female flowers thereafter. (Common sedge has one or two purely male spikes above up to four purely female spikes, sometimes with one transitional spike that is female towards the base and male towards the tip.) Its stems are 30–70 cm (1.0–2.5 ft) tall, with flower-spikes 7–15 mm (0.3–0.6 in) long.

Club sedge grows around the edges of lakes in areas that are under water at some points during the year. It is found across eastern Europe and Scandinavia, and in parts of Asia, as well as across much of North America. It was present on an island in Lough Neagh, Northern Ireland, until it died out in 1886 due to drainage and grazing. In Britain, club sedge is restricted to four sites in Scotland. The first location was discovered in 1895 near Arisaig, and a larger population was later found at Balnagrantach near Inverness. More recently, populations were discovered in Argyll in 1986, and at a second site near Arisaig in 1989. Each of these four populations varies in number from a few hundred to about 1,500, or up to 4,000 in the case of the largest population.

Little research has been carried out into *Carex buxbaumii* in Britain, so the reasons for its scarcity are not well understood. It is reported to be thriving in the sites where it occurs, and its patchy distribution suggests that other populations could yet come to light.

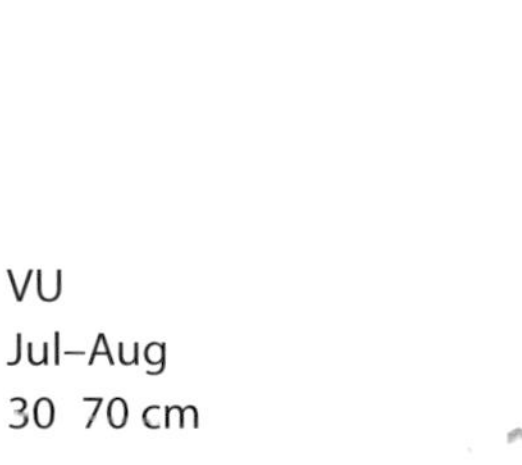

CONSERVATION STATUS	VU
FLOWERING	Jul–Aug
PLANT HEIGHT	30 70 cm

Flowering spikes. (CJD)

Seasonally variable lake at Balnagrantach, Inverness-shire. (CJD)

Starved wood-sedge

Carex depauperata

Starved wood-sedge is a rare and relatively unassuming sedge of deciduous woodlands. It occurs from Ireland south to Spain and across Europe to the Middle East, but appears to be very patchily distributed throughout that area; it is extinct throughout Germany, for instance. In Britain, it has been recorded at around a dozen sites, but only survives at two; a single site is known in County Cork, Ireland.

Starved wood-sedge forms large evergreen tussocks that produce flowering stems up to a metre tall. In this respect, it resembles the very common wood sedge, *Carex sylvatica*, except that its female flower-spikes only produce up to six rather large fruits each. Its leaf-sheaths are flushed red, somewhat like those of the spiked sedge, *Carex spicata*.

Starved wood-sedge was included in Plantlife's *Back from the Brink* campaign, and at one point it was indeed on the brink of extinction in Britain. By the 1950s, it was reduced to two sites in southern England. At one of these (near Axbridge, Somerset) there was just a single plant, and at the other (near Godalming, Surrey) the species had been lost by the early 1970s. At the Somerset site, one man, R. S. Cropper, took it upon himself to monitor and restock the last remaining population, and seemed to have single-handedly kept the species alive in Britain for some time. Meanwhile, after an absence of 15 years, starved wood-sedge made a reappearance in Surrey, after the 'Great Storm' of 1987 felled a large branch from a lime tree at the site where the plant had last been seen. It appears that the disturbance allowed the species to germinate from the seed bank. To supplement that natural population, plants have also been introduced to woodland in the grounds of Charterhouse School, just across the River Wey.

CONSERVATION STATUS EN
FLOWERING May–Jun
PLANT HEIGHT 30–80 cm

Tussock of starved wood-sedge. (CJD)

Shady woodland margin near Godalming. (CJD)

Large yellow sedge

Carex flava

The large yellow sedge is part of a complex group of yellow sedges, and grows 20–70 cm (8–28 in) tall. Three other species (previously treated as subspecies of a broader single species), all quite similar and rather variable, occur in Britain, as do some hybrids between them. There is accordingly some disagreement as to which species occur where, but there is no doubt that the large yellow sedge is the rarest. The combination of a bent beak on the tiny fruits and a stalkless male flower-spike distinguishes this species from its closest relatives.

On a global map, the near-absence of this species from the British Isles is hard to explain; it is widely distributed in neighbouring countries from Portugal to Norway, and over most of continental Europe, as well as in north-eastern and north-western North America. In Britain, though, it is largely or entirely restricted to a single low-lying site on the southern boundary of the Lake District National Park, where it was discovered by the eyebright specialist Dennis Lumb in 1913. A second population, near Malham Tarn in the Yorkshire Dales, belongs either to this species or to the hybrid between it and the common yellow sedge, *Carex demissa*. Other apparently hybrid populations are known, even where the large yellow sedge is not known to occur, leading to the speculation that it may have been more widespread in the past. It may even have been driven out of some sites by the hybridisation itself: when genes can pass between species, the more abundant species can swamp the DNA of its rarer relative and drive it to extinction.

The one certain population of the large yellow sedge is at the boundary between woodland over calcareous rocks and an open fen habitat, a few metres above the nearby Leven Estuary. The putative second site is an open moorside, also over calcareous rocks, at an altitude of around 380 m (1,250 ft).

ALSO IN THE AREA

The limestones of the south-eastern Lake District and north-eastern Lancashire are home to the endemic Lancaster whitebeam, *Sorbus lancastriensis*.

CONSERVATION STATUS	VU
FLOWERING	Jun
PLANT HEIGHT	20–70 cm

Inflorescence. (Andrea Moro; IC)

Roudsea Wood. (CJD)

Bristle sedge

Carex microglochin

Even for a sedge, the bristle sedge is inconspicuous. It grows only a few inches tall, and produces an inflorescence comprising a small number of dull green or brown flowers. It is similar in overall appearance to the much commoner flea sedge, *Carex pulicaris*, or the mountain-dwelling few-flowered sedge, *Carex pauciflora*, all of which have flowers that bend downwards from the stem. The bristle sedge differs in having a distinct bristle extending through the tip of the fruit, but even this resembles the style (part of the female reproductive system) that persists at the tip of the fruit of the few-flowered sedge.

Bristle sedge is an arctic–alpine species: it is found in Arctic and sub-Arctic areas around the world, and also in the Alps, Caucasus, Rocky Mountains and the mountains of Asia. Populations in southern South America are sometimes considered a separate subspecies. In Britain, it is only found as a number of colonies on the slopes of Ben Lawers in the central Highlands of Scotland, at altitudes of 610–945 m (2,000–3,100 ft), where it was first discovered in 1923 by the pioneering aviator Gertrude Bacon and Lady Joanna Charlotte Davy. Although it is only found in a small area, it is doing well under close monitoring, and the number of plants is increasing.

ALSO IN THE AREA

Ben Lawers is home to a number of rare and range-restricted plant species, including snow pearlwort (*Sagina nivalis*), drooping saxifrage (*Saxifraga cernua*, p. 104), scorched Alpine sedge (*Carex atrofusca*), two-flowered rush (*Juncus biglumis*) and chestnut rush (*Juncus castaneus*), as well as other arctic–alpine species with wider distributions in the British Isles.

CONSERVATION STATUS	VU
FLOWERING	Jul
PLANT HEIGHT	5–12 cm

Inflorescence of a dried specimen. (Andrea Moro; IC)

Ben Lawers. (Peter; GG)

Perennial centaury

Centaurium scilloides

Of the five species of centaury native to Britain, the perennial centaury is the only long-lived one, which can be recognised by the fact that it produces non-flowering side-stems. Its flowers are also larger than those of its relatives, at 12–18 mm (0.5–0.7 in) across, making it overall the easiest of the five to identify.

Perennial centaury has a strongly oceanic distribution, being found on all the islands of the Azores and at scattered locations along the coasts of Portugal, northern Spain, northern France and south-western Britain, as well as at sites some distance inland on the continent. In Britain, it occurs in the wild along 3 km of coast near Newport, Pembrokeshire, where it was first discovered in 1918. In the 1950s and 1960s, it also grew at two sites in Cornwall. It was then not seen there for decades until it was rediscovered in 2010, flowering profusely at Porthgwarra near Land's End. The species has also been cultivated in Britain since 1777, and has occasionally escaped from gardens into adjacent verges at several sites in south-eastern England.

The distribution of the perennial centaury seems to be restricted to areas with mild winters, which makes sense for a perennial herb: annual plants can overwinter as seeds, but perennials must survive as fully grown plants. It is also self-compatible, meaning that it can set seed even when there is only a single plant, and it can produce large numbers of seeds. Experiments have shown that these seeds remain buoyant and viable after spending several weeks in sea-water. This, together with the prevailing currents, makes it possible that the coasts of Britain and northern Europe were all colonised by sea from seeds produced by plants much further south.

ALSO IN THE AREA

A dwarf, coastal subspecies of the common juniper, *Juniperus communis* subsp. *hemisphaerica*, is found in Cornwall and possibly in Pembrokeshire (although the latter population may be an extreme form of the widespread subspecies).

CONSERVATION STATUS	EN
FLOWERING	Jul–Aug
PLANT HEIGHT	10–30 cm

Perennial centaury. (CJD)

Coastal grassland near Porthgwarra. (CJD)

Slender centaury

Centaurium tenuiflorum

Slender centaury is an annual species of centaury (unlike the perennial centaury, *Centaurium scilloides*, p. 26) that grows in damp sites near the sea, from Britain to the Mediterranean, and east as far as Pakistan. It has also been introduced to North America, New Zealand and Australia, including Lord Howe Island and Norfolk Island. It is very closely related to the lesser centaury, *Centaurium pulchellum*, but has more leaves along its stems, and holds its branches more upright. It grows up to 30 cm (12 in) tall, with flowers 4–8 mm (0.15–0.30 in) in diameter.

The slender centaury was first found in Britain on the Isle of Wight in 1878, and also once lived in the Channel Islands, but it has since died out in both places (a report of its rediscovery on the Isle of Wight in 1991 is now thought to be a misidentification of the lesser centaury). It is now found in Britain only at seven locations on a short stretch of the coast west of Bridport, Dorset, where it lives on landslips from the unstable sandstone cliffs, and on the undercliffs. The disturbance caused by the fragile nature of the cliffs appears to be necessary for the plant to persist here, and within 10 years of a landslip, a population will die out if there have been no more collapses in the interim.

Most centauries occasionally produce white flowers instead of the usual pink. In Dorset, the slender centaury reverses this trend: the vast majority of plants are white-flowering, with only occasional pale pink flowers. The population that grew on the Isle of Wight until 1953 had mostly pink flowers, in line with populations elsewhere in the world.

CONSERVATION STATUS	VU
FLOWERING	Jun–Sep
PLANT HEIGHT	10–35 cm

Slender centaury growing in Dorset. (CJD)

Golden Cap, and the unstable cliffs where slender centaury grows. (CJD)

Red helleborine

Cephalanthera rubra

Red helleborine is a distinctive orchid with pink flowers, growing 20–50 cm (8–20 in) tall. It is related to two other helleborine species in Britain, both of which have white or off-white flowers. The entirely saprophytic (using fungal decay rather than sunlight for nutrition) phantom orchid, *Cephalanthera austiniae*, is a close relative that grows in North America; it is all white and has no leaves, much like our ghost orchid, *Epipogium aphyllum* (p. 52).

Red helleborine is widespread across the more continental parts of Europe, but is limited in Britain to three populations in southern England. All three sites are in beech woodland, the first on the western edge of the Cotswolds in Gloucestershire (first recorded in 1797, and for a long time the only population known), a second on a steep wooded slope ('hanger') near Hawkley, Hampshire, and a third in the Buckinghamshire Chilterns. These three populations have been compared genetically, and found to be distinct, suggesting that each is the result of a separate migration from the continent following the retreat of the Pleistocene ice sheet.

The red helleborine does not flower every year, and may not even produce leaves. It is a hemisaprophyte, obtaining much of its nutrition from fungi in the soil. In the long term, though, it does still need sunlight, and prefers moderately open ground to thrive. Careful management has allowed the Gloucestershire population to recover from three to nearly 40 shoots. In contrast, too much vegetation was once removed from the Hampshire site, and the red helleborine has not been seen there since 2008, although it might reappear in the future if it can regrow from surviving rootstock.

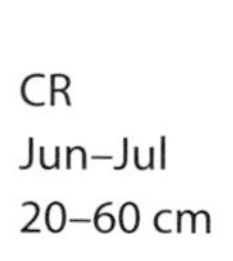

CONSERVATION STATUS CR
FLOWERING Jun–Jul
PLANT HEIGHT 20–60 cm

Red helleborine growing in western France. (Liliane Roubaudi; TB)

Beech woodland in the Gloucestershire Cotswolds. (CJD)

Wood calamint

Clinopodium menthifolium

Wood calamint is a herb in the mint family, growing up to 60 cm (2 ft) tall and with flowers 15–22 mm (0.6–0.9 in) long. It is most similar to the common calamint, *Clinopodium ascendens*, but has larger flowers and leaves. It grows across much of Europe and the wider Mediterranean Region, although its full distribution is clouded by taxonomic and nomenclatural confusion. (The common calamint was formerly treated as a subspecies of *C. menthifolium*, and it is not always clear which of the two species older records refer to.)

In Britain, wood calamint grows only in one dry chalk valley to the west of Newport, Isle of Wight. It has been known at that site since 1843, when it was found 'in the greatest profusion and luxuriance' by the local botanist William Bromfield, who described it as a new species, *Calamintha sylvatica*. (This name is still sometimes seen, although it is no longer correct, since it was recognised that the species had already been described from what is now Slovenia by the Austrian botanist Nicolaus Host.) By the time of his death in 1851, Bromfield had spent 14 years writing a flora of the island, *Flora Vectensis*, which was published in 1856; his flora of Hampshire remains unpublished.

At the time of its discovery, wood calamint was present across several hectares of the valley, but changes in land-management, especially the abandonment of coppicing, severely affected the wood calamint population. In 1960, the road was improved to aid the construction of a television transmitter nearby, and the species was reduced to five clumps. That year, the Isle of Wight Natural History and Archaeological Society took on the species' conservation, and it has since recovered to around 2,000 plants in several patches. There are no past records of wood calamint from any other British sites, suggesting that it has always been restricted to this small area.

ALSO IN THE AREA

Martin's ramping-fumitory (*Fumaria reuteri*) grows in Britain only on the Isle of Wight and in western Cornwall. In both areas, it grows chiefly as a weed of allotments.

CONSERVATION STATUS CR
FLOWERING Aug–Sep
PLANT HEIGHT 30–60 cm

Wood calamint growing on the Isle of Wight. (CJD)

Conservation verge at the woodland's edge. (CJD)

Lundy cabbage

Coincya wrightii

The Lundy cabbage, *Coincya wrightii* (formerly *Rhynchosinapis wrightii*) is a long-lived maritime plant in the cabbage, mustard and cress family. It is classified in the same genus as a number of Mediterranean species, as well as *Coincya monensis*, which is widespread across western Europe, but includes one subspecies (Isle of Man cabbage) that is restricted to the coasts of western Britain and the Isle of Man. Lundy cabbage has a similar habitat to the Isle of Man cabbage but is, as its name suggests, endemic to the island of Lundy in the Bristol Channel, and is thus one of Britain's few truly endemic species of vascular plant.

Lundy cabbage grows up to 130 cm high (4 ft 4 in), with leaves up to 45 cm (18 in) long and divided into lobes, especially towards the base. It produces relatively few flowers, but these are among the largest flowers in the family, and are a bright, cheerful yellow. The fruits are long, fairly straight pods up to 8 cm (3 in) long. It is a much hairier plant than the Isle of Man cabbage, and has leaves with the last lobe much bigger than the side-lobes, rather than all of a similar size.

Lundy cabbage grows along the east coast of Lundy, especially in the southern half of the island. It grows on the cliffs and 'sidelands' (steep slopes descending from the main island plateau), mostly in open situations, and especially where it cannot be easily reached by grazing animals such as sheep and goats. It is outcompeted by the purple-flowering shrub *Rhododendron ponticum*, which has spread over much of the island's coastline, although efforts to remove the *Rhododendron* are helping the Lundy cabbage to thrive. Its population has varied considerably in recent years, with up to 10,000 individuals present in a good year.

Although several vascular plants are endemic to Britain, the Lundy cabbage is the only one known to host an endemic herbivore: the bronze Lundy cabbage flea beetle, *Psylliodes luridipennis*, only feeds on Lundy cabbage.

CONSERVATION STATUS VU
FLOWERING Jun–Aug
PLANT HEIGHT 50–100 cm

Local varieties of several other insects are also restricted to this host-plant. It was actually the unusual insects that first led the entomologist F. R. Elliston-Wright to suspect that the Lundy cabbage was not the same as the Isle of Man cabbage. How its ancestors arrived on Lundy remains something of a mystery; the most similar-looking plants occur in north-western Spain, so the Lundy cabbage's ancestors may have been transported from there shortly after the last ice age.

Lundy cabbage, with flea beetles. (Alan Rowland)

Sidelands of Lundy. (Alan Rowland)

Strapwort

Corrigiola litoralis

Strapwort, *Corrigiola litoralis* (often mis-spelt '*littoralis*'), is a sprawling annual plant of periodically flooded shingle banks. It was formerly assigned to the family Molluginaceae, but is now classified in the family Caryophyllaceae, alongside the pinks, campions, stitchworts and sandworts, on the basis of DNA evidence.

Strapwort is the only British member of its family to have lower leaves that are alternate (rather than paired) and with stipules at their bases. Each leaf is up to 4 cm (1.5 in) long and 5 mm (0.2 in) wide, and fairly fleshy. From June to August, it produces white flowers, each about 2 mm wide, in rounded clusters.

Strapwort grows wild around the Mediterranean, and inland in much of western Europe, western Asia and Africa; it is reported to be an abundant weed of Kenyan coffee plantations above 2,500 m (8,200 ft). It was first found in Britain by William Hudson in the 1780s, too late for him to include it in his *Flora Anglica*. The British distribution is now limited to the population that Hudson found at Slapton Ley in Devon, after another at The Loe in Cornwall died out in the 1930s. Both sites feature large freshwater lakes separated from the sea by a narrow shingle spit. The suggestion that strapwort is native to Britain has been doubted, but the discovery of its seeds in mid-Pleistocene natural deposits in Herefordshire argues strongly against it being simply an introduction. Strapwort needs open ground and is outcompeted by other plants such as knotweed (*Polygonum aviculare*) that grow in similar locations. It therefore does best when there is a certain amount of disturbance, which creates new openings and removes potential competitors. This disturbance is normally achieved through a mixture of grazing by cattle, trampling by people, and fluctuating water levels.

Conservation measures have been introduced at Slapton Ley to remove competing vegetation and expand the area available to strapwort. As a result, the number of individuals had increased to nearly 2,000 plants by 2010. It is also being re-introduced to The Loe, and if successful, the programme may be expanded to other sites.

CONSERVATION STATUS	CR
FLOWERING	Jul–Aug
PLANT HEIGHT	5–25 cm

Growing in north-western Germany. (Christian Fischer; WC)

View over Slapton Ley. (CJD)

Pigmyweed

Crassula aquatica

The name 'pigmyweed' gives some indication of the size of this plant, which has been rightly described as 'easily overlooked'. The whole plant is usually less than 5 cm (2 in) tall, with narrow leaves up to 7 mm (¼ in) long, and tiny (1–2 mm) flowers borne singly in the axils of the leaves. Three other species of *Crassula* occur in Britain, only one of which, mossy stonecrop (*Crassula tillaea*), is native. Pigmyweed differs from mossy stonecrop in having four petals and stamens in each flower, rather than three. The two exotic species both have flowers on short stalks.

Pigmyweed grows beside and beneath fresh water, flowering in August and apparently self-fertilising. The greatest threat to its survival is eutrophication – an excess of nutrients, such as through the use of fertilisers. When nutrient levels increase, the habitat becomes amenable to plants such as reeds, which can outcompete and smother the pigmyweed.

Pigmyweed has a fairly broad but extremely patchy distribution in the Northern Hemisphere. In Europe, it is most abundant in the countries around the Baltic Sea. In Britain, it was first discovered at Adel Dam on the outskirts of Leeds in 1921, but has not been seen there since 1938. A second population was discovered in 1969 over 250 miles away, on the River Shiel on Scotland's west coast. This is an unlikely site for a deliberate introduction, so it is most likely that this population arose naturally. Given the species' inconspicuous nature, the fact that it can complete its whole life cycle under water, and the vast areas available in western Scotland, it is entirely possible that there are other populations of pigmyweed in Britain that have not yet been recognised.

ALSO IN THE AREA

The mountain ridge where diapensia (*Diapensia lapponica*, p. 48) grows is only 25 km (16 miles) as the crow flies from the part of the River Shiel where pigmyweed grows.

CONSERVATION STATUS	VU
FLOWERING	Jun–Jul
PLANT HEIGHT	2–5 cm

Pigmyweed, Texas. (Layla Dishman)

River Shiel at its exit from Loch Shiel. (Trevor Littlewood; GG)

Leafless hawk's-beard

Crepis praemorsa

The yellow-flowering composites of the tribe Cichorieae - including the hawk's-beards, hawkweeds, hawkbits, dandelions and sow-thistles - are a difficult group for identification. (Mediaeval people thought that hawks ate these plants to improve their eyesight, which explains the abundance of hawks in the common names.) Among the 10 species of hawk's-beard that grow in Britain, leafless hawk's-beard is the easiest to identify, as it has no leaves on the stem, which rises up to 60 cm (2 ft) from a rosette of leaves at ground level. Several members of other genera in the same tribe share this appearance, though, and proper identification involves details of the flower-heads and seeds.

Leafless hawk's-beard is found chiefly in eastern Europe and southern Scandinavia, and some believe it to be an introduction in Britain. It was found in 1988 in eastern Cumbria, growing near a stream in a shallow-soiled, lightly grazed, unimproved hay-meadow over limestone at an altitude of around 240 m (790 ft). The site is largely undisturbed and very similar to the situations in which the leafless hawk's-beard occurs in Scandinavia, all of which suggests that its occurrence there is natural, rather than a deliberate or accidental introduction.

Despite regularly producing flowers, the British population of leafless hawk's-beard does not appear to produce any viable seed. This suggests that it is a single clone, reproducing vegetatively rather than sexually, and so unable to colonise new areas away from the existing population. Eleven patches are known at the one site, with a total of around 200 individual plants.

ALSO IN THE AREA

Alchemilla minima, the least lady's-mantle, is endemic to Ingleborough and Whernside in the Yorkshire Dales, not far to the south. *Bartsia alpina*, the Alpine bartsia, occurs on the Orton Scar, in Upper Teesdale, near Malham Tarn in the Yorkshire Dales, and in a small area of the central Scottish Highlands.

CONSERVATION STATUS	EN
FLOWERING	May–Jun
PLANT HEIGHT	20–60 cm

Leafless hawk's-beard growing in Austria. (Hermann Schachner; WC)

Unimproved hay-meadow in eastern Cumbria. (CJD)

Lady's-slipper orchid

Cypripedium calceolus

The lady's-slipper orchid is unmistakable when in flower, with a swollen yellow lip surrounded by four burgundy-coloured tepals up to 5cm (2in) long.

The lady's-slipper orchid can reasonably be described as Britain's rarest plant. Although it was once found across a modest area of northern England, the wild population has for decades been reduced to a single individual. The precise location of that plant is a closely guarded secret, but it is known to be somewhere near Malham in the Yorkshire Dales. The factors stopping the lady's-slipper orchid from expanding its population may not be well understood, but it is all too clear what reduced it to this situation. Orchids have always attracted amateur botanists, and the lady's-slipper orchid has by far the showiest flowers of any British orchid, and was well known to early botanists (the first British record is from 1633). As a result, it was decimated by collectors, especially in Victorian times. There are records of it being collected as late as the 1950s, by which time it was obvious that the species was at severe risk of being completely wiped out in Britain. Indeed, notes accompanying herbarium specimens show that local botanists were aware of the harm being caused and, as early as the 1820s, were reluctant to reveal exactly where the plant grew. Their reticence has proven to be well-founded.

Outside Britain, the lady's-slipper orchid is widespread, although never common, across much of Europe and temperate Asia, and a close relative occurs widely in North America. A further 120 species of slipper orchid are found across the Northern Hemisphere, mostly on calcareous soils.

A concerted programme of cultivation and re-introduction is under way in northern England. Individuals have been planted out at 12 widely scattered sites. The British plants do not appear to differ genetically from their continental relatives, and a surprisingly large amount of genetic diversity has been retained in the few plants in cultivation, all of which bodes well for the species' future in Britain.

CONSERVATION STATUS	CR
FLOWERING	May–Jun
PLANT HEIGHT	20–60 cm

ALSO IN THE AREA

The English sandwort, *Arenaria norvegica* subsp. *anglica*, is endemic to limestone pavements to the north of Malham.

Lady's-slipper orchid growing in the Dolomites. (CJD)

Deciduous woodland over limestone near Malham. (CJD)

Starfruit

Damasonium alisma

Starfruit is a water plant that grows in western Europe. It should not be confused with the tropical tree known as starfruit or carambola, *Averrhoa carambola*, although both are aptly named for their star-shaped fruits. It is similar to the much commoner water-plantain, *Alisma plantago-aquatica*, although in that genus the lobes of the fruit are strongly curved inwards. Starfruit can grow up to 60 cm (2 ft) tall, with flowers 5–9 mm (0.2–0.4 in) across.

Throughout its range, starfruit is dangerously rare, but nowhere more so than in Britain. From a total of 100 sites where the species had been recorded in Britain, it dwindled to a single small population in Buckinghamshire in the 1980s. Its seeds are long-lasting, however, and starfruit has reappeared following clearances at two further sites in the county – in one case, after an absence of nearly 90 years. It has also been planted at additional sites in Buckinghamshire and Surrey.

Starfruit is a demanding plant: it needs bare soil with little competition, it needs its seeds to be covered in water for them to germinate, but also needs the water to recede in order to flower and set seed. As a result, it only persists where water levels fluctuate, and this is the reason for its decline; most ponds are no longer trampled by livestock, and their levels are often carefully stabilised. Villages in south-east England value pristine duckponds more highly than dank, muddy pools, it would seem. Several of the ponds where it had formerly been recorded are now managed specifically to foster the starfruit, by clearing them of vegetation and allowing the water level to vary over the year.

ALSO IN THE AREA

Gentianopsis ciliata, the fringed gentian (p. 62), grows on the other side of the Chilterns.

CONSERVATION STATUS CR
FLOWERING Jun–Aug
PLANT HEIGHT 5–40 cm

Starfruit growing in south-western France. (Florent Beck; TB)

Pond edge within lowland heath. (CJD)

Cheddar pink

Dianthus gratianopolitanus

Cheddar pink is a perennial, mat-forming species of pink, related to carnations and campions. It is found across central Europe, from France to Poland (an outlying population that grew near Zalishchyky on the banks of the River Dniester in western Ukraine has died out). Its English name reflects the fact that the species is restricted in Britain to the area around Cheddar Gorge in the Mendip Hills, whereas its scientific name refers to Grenoble, where it was found by the French botanist who first described it, Dominique Villars.

Cheddar pink grows in crevices and on ledges cutting into the limestone cliffs, and produces bright pink flowers up to 30 mm (1.2 in) across. It differs from the other few-flowered pinks by the shallow pinking of its petals and its entirely hairless stems. The clove pink (*Dianthus caryophyllus*) grown in gardens is similar but has larger flowers.

Cheddar pink was discovered in Britain in 1696 by the young botanist Samuel Brewer, and was at one time collected in large quantities and sold to tourists visiting the gorge. Thankfully, such excesses have since been curbed, but conservation in the area is left in a quandary. To best protect the open grassland that favours the Cheddar pink, encroaching scrub would need to be removed, but that is likely to harm the populations of rare whitebeams, so site managers have to maintain a careful balance between the two habitats.

ALSO IN THE AREA

Cheddar Gorge is home to a number of rare plants, such as the slender bedstraw (*Galium pumilum*), and houses several uncommon species of whitebeam (*Sorbus*), including three species endemic to the gorge. The surrounding Mendip Hills also contain most of the British population of the Somerset hair-grass, *Koeleria vallesiana*.

CONSERVATION STATUS	VU
FLOWERING	Jun–Jul
PLANT HEIGHT	10–20 cm

Cheddar pink growing in the Vallée d'Aosta, Italy. (Andrea Moro; IC)

Cheddar Gorge and surroundings. (Chris Andrews; GG)

Diapensia

Diapensia lapponica

Diapensia lapponica is a plant with no real English-language vernacular name. It is generally referred to by the generic name *Diapensia*, a corrupted form of a classical name that originally signified the sanicle - a plant with which it shares nothing in particular. Previously considered a single species with two subspecies, diapensia has recently been divided into two species - *Diapensia obovata* on either side of the Bering Strait, and *Diapensia lapponica* stretching from the central Canadian Arctic across the North Atlantic to the Yenisei River in northern Siberia. Its British occurrences are the southernmost outposts of the genus in Europe.

Diapensia is in many ways a typical Arctic plant; it grows as a mat or cushion of leaf rosettes packed densely together. Its flowers are large compared to the leaves - 15–20 mm (0.6–0.8 in) in diameter - and are white, presumably to attract a wide range of generalist pollinators. It is only likely to be confused with some saxifrages when in flower (although they have 10 stamens per flower and two-lobed stigmas).

In Britain, diapensia is found only on two relatively unremarkable mountains in the western Highlands of Scotland. The first has twin summits, Sgùrr an Utha (796 m; 2,612 ft) and Fraoch-bheinn (790 m; 2,592 ft), connected by a broad ridge that drops to around 735 m (2,410 ft) over its 1-km length and forms the habitat for diapensia. Because these summits are a long way off the beaten track, diapensia was not recorded here until 1951. The rock is ancient, similar to that of the surrounding mountains, many of which are considerably higher. Although only a single cushion was initially discovered, around 1,200 are now known. The second site, around 15 miles north near Loch Quoich, was discovered 25 years later, and no further populations have been found since, despite the likelihood that others may exist between the two.

The sites where diapensia grows, both in Britain and abroad, are characterised by having few nutrients available, but being severely exposed

CONSERVATION STATUS	VU
FLOWERING	Jul
PLANT HEIGHT	3–8 cm

to winds. It has been suggested that the plant derives much of its mineral needs from particles carried on the wind.

ALSO IN THE AREA

The riverbank where pigmyweed (*Crassula aquatica*, p. 38) grows is only 25 km (16 miles) as the crow flies from Sgùrr an Utha and Fraoch-bheinn.

Diapensia growing in western Greenland. (Kim Hansen; WC)

Wind-exposed rocks on Fraoch-bheinn. (CJD)

Yellow whitlow-grass

Draba aizoides

Yellow whitlow-grass is an early-flowering cushion-plant that grows in limestone crevices. It is quite distinctive, with its groups of yellow flowers on stems up to 10 cm (4 in) high above rosettes of simple, pointed, bristle-edged leaves. It is found from the mountains of northern Spain to the Carpathians, including the Pyrenees, Alps, Jura, Balkans and the limestone mountains of southern Germany. Its occurrence in Britain is therefore something of an anomaly, being far to the north-west of an otherwise fairly contiguous range, although another isolated population occurs on cliffs beside the River Meuse (or Maas) in Belgium.

Yellow whitlow-grass was reportedly discovered in Britain in 1795 by John Lucas the younger, although he didn't mention it to his friend William Turton until after Turton had found it for himself at Pennard Castle. It occurs along two sections of limestone cliff on the southern edge of the Gower Peninsula, separated by the sandy expanse of Oxwich Burrows. Collectively, they span 17 km (10 miles), from Pwll Du Head in the east to Tears Point in the west. This line of cliffs is also where the so-called Red Lady of Paviland was found – a 33,000-year-old red-dyed skeleton. At the time he was alive (the epithet 'Lady' proved to be an error), most of the British Isles were covered in a thick ice-sheet, but the furthest parts of the Gower Peninsula were spared, making them a potential ice-age refugium – a small area in which animals and plants can survive. The genetic distinctness of the Welsh plants from their continental cousins supports this hypothesis.

The Gower populations of yellow whitlow-grass have been damaged by overzealous collecting, and by the spread of non-native *Cotoneaster* plants over their habitat, but are recovering well, and efforts under way to remove the invasive *Cotoneaster* will surely help.

CONSERVATION STATUS NT
FLOWERING Mar–May
PLANT HEIGHT 5–15 cm

ALSO IN THE AREA

Sea stock, *Matthiola sinuata*, is found on mobile dunes in western Glamorgan and along a short stretch of coast on the other side of the Bristol Channel, but nowhere else in Britain. The dune gentian, *Gentianella uliginosa*, grows on dunes in Glamorgan and Devon, and perhaps on the Scottish island of Colonsay.

Yellow whitlow-grass growing in the French Alps. (Jean-Luc Gorremans; TB)

Pennard Castle and nearby cliffs. (CJD)

Ghost orchid

Epipogium aphyllum

All orchids rely on soil fungi for at least part of their life cycle, but some live entirely as parasites on fungi. These 'saprophytic' plants, including the ghost orchid, have no need for chlorophyll, and so lack leaves or any green colouring. Because they are not dependent on sunlight, they can occur in the darkest parts of forests, and are generally impossible to locate except when they are flowering. Other examples include the bird's-nest orchid (*Neottia nidus-avis*) and the coralroot orchid (*Corallorrhiza trifida*). The yellow bird's-nest or dutchman's pipe (*Hypopitys monotropa* or *Monotropa hypopitys*) is similar in appearance and ecology, but belongs to the heather family, Ericaceae.

The ghost orchid's flower spikes are up to 25 cm (10 in) tall, but typically much less, and often hidden by leaf litter. They are pale brownish-pink, with up to four pale pink flowers, each 15–20 mm (0.6–0.8 in) tall. The flowers are quite unlike those of the other saprophytic orchids in Britain, making the plant unmistakable. It is found in mountainous areas of continental Europe, and across north-eastern Europe, with scattered occurrences in a narrow band across Asia. The British sites are the westernmost on record.

The ghost orchid occurs chiefly in dense, ancient beech woodland, but has also been found in oak woods. Despite its distinctiveness, it is 'the most difficult British plant to find', and was not discovered in Britain until 1854, when a Mrs Anderton Smith found one growing in woodland near Tedstone Delamere, Herefordshire. The woodland was felled later that year. Ghost orchids were found at a second site, in Bringewood Chase, in 1876, 1878 and 1892; that woodland was felled during the First World War. A site near Ross-on-Wye was discovered in 1910, and thankfully remains undamaged. Between the two world wars, the ghost orchid was repeatedly found by schoolgirls in the Chilterns. In 1924, a Miss Butler found it west of Henley-on-Thames, and in 1931, Miss Vera Smith found it four miles to the south-west. Woods near Marlow produced a fine display of flowering spikes in 1953, and continued to produce finds until 1987.

CONSERVATION STATUS	EX
FLOWERING	Jun–Sep
PLANT HEIGHT	10–30 cm

For some years after, there were no British sightings and, by the early 2000s, it was declared to have died out. This proved premature when one was found in Herefordshire in 2009 – the first in western England for over 70 years. There have been vague reports of other sightings in the intervening years, but that sighting remains the only certain record in the last 30 years. Given its preternatural ability to pass unseen, though, the ghost orchid may still be lurking underground elsewhere, flowering surreptitiously among the fallen leaves.

Flowering spikes in southern Germany. (Hans Stieglitz; WC)

Beech woodland near Satwell. (Des Blenkinsopp; GG)

Irish spurge

Euphorbia hyberna

Most of the plants that are rare in Britain are more common in neighbouring areas, generally in continental Europe. Irish spurge is an exception, in that - as its name suggests - it is native to Ireland, where it is widespread across the old province of Munster, and also occurs in a small area of northern County Donegal. It also grows in Atlantic parts of continental Europe, as far as Portugal, and is thus part of the so-called Lusitanian biota, which includes a number of species found in south-western Ireland and, disjunctly, in the Iberian Peninsula. (Other examples include the Kerry slug, *Geomalacus maculosus*, and the strawberry tree, *Arbutus unedo*.)

Within Britain, Irish spurge is found in two small areas of south-western England. One is in the Lyn valley on the Devon–Somerset border, where it was discovered in 1840. The other is at Nance Wood near Portreath, Cornwall, where it was found in 1883. In both cases, it grows in the dappled sunlight of old woodlands and flowers before the trees are fully in leaf - although in Ireland it also occurs along hedge-banks and other grassy areas outside woods. It is long-lived, but is threatened by the spread of Japanese knotweed (*Fallopia japonica*) at some of its sites.

Spurges have bizarre floral structures called cyathia that are hard to reconcile with other flowers. Irish spurge is a medium-sized spurge, up to 60 cm (2 ft) tall, with four fat, sausage-shaped glands encircling the cyathium, and long papillae on the surface of the fruit. The only other British spurges with similar characteristics are the rare, straggling annual *Euphorbia stricta* (upright spurge) and the sweet spurge, *Euphorbia dulcis*, which has leaf-green rather than yellowish bracts.

ALSO IN THE AREA

The Exmoor coast around the Lyn valley is also home to three endemic whitebeam micro-species: *Sorbus vexans* (bloody whitebeam), *Sorbus subcuneata* (Somerset whitebeam) and *Sorbus admonitor* (no-parking whitebeam).

CONSERVATION STATUS	VU
FLOWERING	May–Jul
PLANT HEIGHT	30–60 cm

Irish spurge growing in the French Pyrenees. (CJD)

Dappled shade in Nance Wood, Portreath. (CJD)

Radnor lily

Gagea bohemica

Radnor lily is a yellow-flowering perennial plant that grows from bulbs, very similar to the more widespread yellow star-of-Bethlehem, *Gagea lutea*, but only 1–4 cm high (0.4–1.6 in) and with a larger number of shorter and very narrow leaves. The yellow petals are 12–18 mm (0.50–0.75 in) long and turn white when they age. As a result, when a single shrivelled specimen was first discovered in Britain late in the spring of 1975, the plant was thought to be the Snowdon lily, *Gagea serotina* (p. 58). The hairy stem, and the later discovery that the flowers are bright yellow showed that it could not be Snowdon lily, and it was identified as *Gagea bohemica*, a species found from Central Europe to south-western Asia.

In Britain, Radnor lily is only found at Stanner Rocks, just inside the Welsh border near Radnor. On the continent, it is a species of steppes, and its habitat at Stanner Rocks is similar in many ways; it grows in shallow soils on steep south- and east-facing slopes that dry out in summer. The rain shadow of the Cambrian Mountains must also lend the site a more continental climate than many other sites in Britain, which may help to explain the prevalence of rare plants there.

ALSO IN THE AREA

Stanner Rocks is home to a number of rarities, including the sticky catchfly (*Silene viscaria*; formerly *Lychnis viscaria*), perennial knawel (*Scleranthus perennis*, p. 112), upright clover (*Trifolium strictum*, p. 130) and spiked speedwell, *Veronica spicata*, as well as the oldest rocks in Wales.

CONSERVATION STATUS	VU
FLOWERING	Feb–Mar
PLANT HEIGHT	2–5 cm

Radnor lily growing in Austria. (Stefan.lefnaer; WC)

Stanner Rocks. (Andrew Dixon)

Snowdon lily

Gagea serotina

The Snowdon lily was first collected by the botanist Edward Lhuyd (Llwyd/ Lloyd) in the 17th century, and was for many years named *Lloydia* in his honour. It is now known to belong within the genus *Gagea* (named after the botanical aristocrat Thomas Gage, 1781–1820), which also contains the yellow star-of-Bethlehem and the Radnor lily (*Gagea bohemica*, p. 56). Snowdon lily grows up to 15 cm (6 in) high, and differs from most other *Gagea* species in having white rather than yellow flowers – although a yellow-flowering variety occurs on the islands of British Columbia, western Canada. The tepals are 9–12 mm (0.4–0.5 in) long and often have purplish veins on the outside.

Snowdon lily occurs in the mountains of central and southern Europe, and across Asia, into Alaska, and along the chain of the Rocky Mountains as far south as New Mexico. The five British populations are all found in a small area of northern Snowdonia, and are genetically distinct from both the Alpine and the North American populations. The Welsh plants are only found in steep and inaccessible areas, which protects them from grazing by sheep, although the imposing landscape also draws rock-climbers. Measures are in place to limit damage caused by climbers, but climate change may pose an even greater threat, as the cliffs' height limits the Snowdon lily's ability to survive under a warming climate – there may simply be no higher ground that it can retreat to.

ALSO IN THE AREA

Hieracium snowdoniense is a hawkweed microspecies endemic to Snowdonia, and the shores of the Menai Strait are home to some endemic whitebeam microspecies. The wild cotoneaster on the Great Orme peninsula may be either an endemic species (*Cotoneaster cambricus*) or a garden escape belonging to a common European species; opinions differ.

CONSERVATION STATUS	VU
FLOWERING	May–Jul
PLANT HEIGHT	5–15 cm

Snowdon lily on the cliffs of Snowdon. (CJD)

Clogwyn Du'r Arddu. (CJD)

Alpine gentian

Gentiana nivalis

Alpine gentian is unusual in that it has both an arctic-alpine distribution and an annual life cycle. Its global distribution consists of a number of separate areas: Canada's east coast (Labrador and just extending into Quebec), Greenland, Iceland, Scandinavia, the Pyrenees, the Alps and Apennines, and the Caucasus. There are also a number of more scattered occurrences, including some in the Carpathians and Balkans, and two in Scotland.

The Scottish populations grow on the Ben Lawers range in the central Highlands, where they were first found by the botanist and mycologist James Dickson in 1792, and on the cliffs surrounding Caenlochan Glen, 40 miles (65 km) to the north-east.

Gentians in the genus *Gentiana* differ from *Gentianella* species by the presence of additional lobes between the main petals, and the Alpine gentian differs from the perennial spring gentian, *Gentiana verna* (the only *Gentiana* species likely to be confused with *G. nivalis* in Britain), by its annual habit and much smaller flowers, only 7–11 mm (0.30–0.45 in) across. Being an annual plant in a montane environment is a risky strategy, and the Alpine gentian overcomes this by having seeds that can remain in the soil for decades before germinating.

Alpine gentian plants are easily damaged by grazing, such as by sheep, deer or hares, but grazing by sheep benefits the population overall, by removing competitors and perhaps through changes to the seed bank. From 1987 to 1996, the Ben Lawers population was protected from grazing by sheep, and numbers declined. After grazing was restored, the gentian population improved again. The flowers require the sun's warmth in order to open, and will close up rapidly if a shadow passes over them.

ALSO IN THE AREA

The area around Caenlochan Glen is the only location in Britain where the Alpine blue sow-thistle, *Cicerbita alpina*, grows.

CONSERVATION STATUS	NT
FLOWERING	Jun–Jul
PLANT HEIGHT	1–15 cm

Alpine gentian growing in the Swiss Alps. (Hans Hillewaert; WC)

The cliffs of Caenlochan Glen from the south. (Iain A. Robertson; GG)

Fringed gentian

Gentianopsis ciliata

Fringed gentian (now *Gentianopsis ciliata*, but sometimes still referred to as *Gentianella ciliata*) is an annual or biennial herb that grows in calcareous meadows across central and eastern Europe, and in some parts of western Asia.

It has four rich violet-blue petals, each 10–18 mm (0.4–0.7 in) long, with very distinctive fringing around the edges, especially near the centre of the flower.

In Britain, it is restricted to one site, where it is presumed to be native. That site is in the Chiltern Hills near Wendover, on a west-facing slope. Despite its striking colour, this species has not only been largely overlooked in Britain, but positively refuted. In 1924, the Oxford botanist George Claridge Druce dismissed a report by a Miss M. Williams dating from 1875, but Williams had pressed a herbarium specimen, and her plant is unmistakably the fringed gentian. It was only when the plant was rediscovered in 1982 that the 1875 record was reassessed. (Elizabeth Grosvenor Hood painted the plant in 1873, but kept her painting between the pages of her copy of George Bentham's *Handbook of the British Flora*, rather than publishing it, perhaps because she was not confident about the identification.) An 1892 specimen from Wiltshire, discovered at the Natural History Museum 100 years later, lends further credence to the native status of this species in Britain.

Only one flower was recorded in 2011, and a project to help the fringed gentian increase its numbers in the Chilterns ran into difficulties when seed stored at the Millennium Seed Bank proved to be inviable. Fringed gentian therefore appears to be on the brink of extinction in Britain.

ALSO IN THE AREA

Both starfruit (*Damasonium alisma*, p. 44) and *Gentianella germanica*, the Chiltern gentian, are hardly found in Britain outside the Chiltern Hills. The area is also home to many orchids, including the military orchid (*Orchis militaris*, p. 80) and monkey orchid (*Orchis simia*, p. 82).

CONSERVATION STATUS	CR
FLOWERING	Aug–Oct
PLANT HEIGHT	5–30 cm

Fringed gentian growing in the Austrian Alps. (CJD)

Chalk grassland on Coombe Hill, Buckinghamshire. (CJD)

Fringed rupturewort

Herniaria ciliolata

Fringed rupturewort is a trailing evergreen plant of open ground near the coast, with leaves 2–6 mm (0.10–0.25 in) long. Three subspecies are recognised, one endemic to Cornwall, one endemic to the Channel Islands, and one found along the coasts of continental Europe from western France to northern Portugal.

Fringed rupturewort is much woodier at the base of its stems than the other two species of rupturewort that occur in Britain. Smooth rupturewort (*Herniaria glabra*) can be somewhat woody, but has more pointed fruits that contain smaller seeds than those of fringed rupturewort. (The third species – hairy rupturewort, *Herniaria hirsuta* – differs from the other two in having hairs not only on the edges of the leaves, but also on the flat surfaces.) The prostrate habit of fringed rupturewort allows its flowers to be reached by the ants that pollinate them.

Within Great Britain, fringed rupturewort is only found on the Lizard Peninsula in Cornwall. Its occurrence there was first noted by the famous botanist John Ray, who visited in 1667. Located at the southernmost tip of the island of Great Britain, the Lizard is a well-known area for rare plants, due in part to the peninsula's mild, maritime climate with hardly any frosts, and in part to the underlying serpentine bedrock. Serpentine is rich in magnesium but contains little calcium; this means that calcium-hating plants, such as those typically found on acid heathland, can occur alongside base-loving plants, typically found in lime-rich areas such as chalk downland. The low nutrient levels also prevent the area from being taken over by more generalist species.

ALSO IN THE AREA

The Lizard Peninsula is home to many rare plants. As well as the clovers (see the twin-headed clover, *Trifolium bocconei*, p. 128, and the upright clover, *Trifolium strictum*, p. 130; also the long-headed clover, *Trifolium incarnatum* subsp. *molinerii*), the pygmy rush (*Juncus pygmaeus*, p. 70) and land quillwort (*Isoetes histrix*, p. 68) are also found nowhere else in Britain.

CONSERVATION STATUS	VU
FLOWERING	Jul–Aug
PLANT HEIGHT	5–30 cm

Fringed rupturewort. (CJD)

Rough ground on the Lizard. (CJD)

Esthwaite waterweed

Hydrilla verticillata

The name 'Esthwaite waterweed' has been applied to two different species of water plant with leaves in whorls. One is *Elodea nuttallii* (more usually called 'Nuttall's waterweed'), which is a widespread non-native plant, originally from North America; the other is a native and very rare species, *Hydrilla verticillata*. Its English name derives from the fact that it was discovered in Esthwaite Water in the Lake District in 1914, although it was last seen there in 1941, and is presumed to have died out through eutrophication. It had been found at a site in County Galway, Ireland, in 1935, and was discovered in 1999 in Bargatton Loch, an otherwise fairly unremarkable lake on private land in Dumfries and Galloway.

The genus *Hydrilla* contains only this one species, and differs from *Elodea* in having two tiny fringed scales at the base of the leaves; in *Elodea*, these scales are either entirely absent or lack the fringes. The leaves of *Hydrilla* are 5–20 mm (0.2–0.8 in) long, it has more leaves in each whorl than *Elodea*, and the minuscule teeth along the edges of the leaves continue all the way to the leaf-bases. *Hydrilla* is dioecious – it has separate male and female plants – but the plants in Britain have never been seen to flower (the Irish plants, by contrast, are female). Instead, the plants reproduce through specialised buds that break off and grow into new plants.

Globally, *Hydrilla* is found across much of Europe and Asia, and in parts of Africa and Australasia. It has been introduced to North America, where it has become an invasive species, particularly in the Atlantic and Gulf states of the USA. It is difficult to distinguish from other waterweeds, and may occur in other unpolluted lakes across Great Britain.

CONSERVATION STATUS	VU
FLOWERING	n/a
PLANT HEIGHT	20–100 cm

A flowering stem out of water. (Big Cypress National Preserve, Florida)

Aquatic vegetation at Bargatton Loch. (CJD)

Land quillwort

Isoetes histrix

Quillworts are a peculiar group of plants related to the clubmosses and perhaps even more closely to *Lepidodendron*, one of the dominant trees in the Carboniferous forests that were later turned into the coal measures. Despite these lofty relatives, modern quillworts are not spectacular plants. They consist of a rosette of leaves arising from a stubby stem with short roots, and reproduce via spores produced in hardly visible structures at the base of the leaves.

Most quillwort species, including two of the three British species, are aquatic, but the land quillwort is helpfully distinct in growing out of water rather than in cold mountain lakes. Although it is therefore easy to tell apart from the other British quillwort species, this does bring it into confusion with a number of unrelated flowering plants that have similar rosettes of narrow leaves, including shoreweed (*Littorella*; Plantaginaceae), awlwort (*Subularia*; Brassicaceae), *Lobelia* (Campanulaceae), spring squill (*Scilla verna*; Asparagaceae), thrift (*Armeria maritima*; Plumbaginaceae) and some grasses, sedges and rushes. The shiny, hairless leaves of land quillwort are typically up to 4cm (1.6in) long and around 1mm wide, and are often spirally curved, although less curved than the tight coils of spring squill.

Land quillwort grows in Mediterranean parts of Europe, and along the Atlantic coast as far as the Channel Islands, reaching its northern limit in Britain. Here, it only grows on the Lizard Peninsula, especially its western aspects – although it was not discovered until 1919, and even then the discoverer, Fred Robinson, suffered the ignominy of having his report discounted by the illustrious George Claridge Druce. It wasn't until 1937 that the record was set straight after Ronald Melville re-found the plant, and sent a specimen back to his colleagues at Kew for confirmation. It grows in fairly bare locations, including areas of rough grazing and moderately trampled footpaths, including the South West Coast Path.

ALSO IN THE AREA

See under fringed rupturewort (*Herniaria ciliolata*, p. 64).

CONSERVATION STATUS	VU
FLOWERING	Apr–May
PLANT HEIGHT	2–10 cm

Land quillwort growing in southern Italy. (Domenico Puntillo; IC)

South West Coast Path on the Lizard Peninsula. (CJD)

Pygmy rush

Juncus pygmaeus

Pygmy rush is a very small plant, described in one assessment as 'a tiny plant, uncharismatic, and probably of little interest to the general public'. Its habitat is also rather unbecoming – it grows in the muddy puddles and ruts that form along cart-tracks and around gateways.

Pygmy rush is one of several small rushes with cylindrical leaves that are buttressed internally, but it is the only British one of them to be annual. The plants are green early in the season, but turn a pinkish red as the ground dries out. They produce groups of 2–3 small, dull flowers on stalks up to 8 cm (3 in) tall.

Pygmy rush occurs across much of western and southern Europe, as well as parts of North Africa and just reaching into Turkey. Within Britain, it is only found on the Lizard Peninsula, where it grows in the tracks that pepper the maritime heathlands there. Its occurrence on the Lizard was discovered in 1872 by W. H. Beeby, a botanist more frequently found at the opposite end of our archipelago, in Shetland.

Changing farming practices mean that tracks on the Lizard are no longer disturbed by the hooves of oxen and wooden cart-wheels as they once were. As a result, the populations of pygmy rush have been struggling to survive. Of the 22 discrete populations known to have existed since 1950, only eight were known to survive by the century's end. Some former earthen tracks have even been filled in with hardcore, destroying the pygmy rush's home, presumably unwittingly. Having recognised the threat to the pygmy rush, and the conditions needed to aid its survival, conservation efforts have been designed to maintain the wet, rutted tracks it requires. This is not an especially pretty species, nor an especially pretty habitat, but conserving them allows at least one more native species to survive.

CONSERVATION STATUS	EN
FLOWERING	May–Jul
PLANT HEIGHT	2–8 cm

ALSO IN THE AREA

Many rare plants grow on the Lizard (see 'Fringed rupturewort', p. 64); the yellow centaury, *Cicendia filiformis*, grows in the same microhabitat as the pygmy rush and is easier to spot, so can be used as an indicator of suitable habitat for pygmy rush.

Pygmy rush turning red as the habitat dries out, Corsica. (Hugues Tinguy; TB)

Farm track through Cornish heathland. (CJD)

Least lettuce

Lactuca saligna

Least lettuce is widely distributed across Europe, western Asia and North Africa. It has been introduced to much of North America, to Australia and New Zealand, and to parts of South America. In Britain, it is near the northern limit of its range, and was once found here at scattered locations from Sussex to Norfolk. Nowadays, it is limited to three or four locations across three counties. In East Sussex, it occurs at six positions in Rye Harbour; in Kent, it occurs on the Isle of Grain and the Isle of Sheppey, either side of the River Medway; in Essex, it occurs at Fobbing, close to the London Gateway port and the Coryton power station and refinery. The Sussex population was flooded with sea-water in the 1990s, causing a severe decline from around 50,000 individuals to around 1,000, but the Essex population is healthy, with around 15,000 individuals, and a population in Kent was rediscovered in 1999.

Least lettuce grows up to 1 m (3 ft) tall. It lives in a variety of open habitats close to the sea, including salt-marshes, sea walls, shingle and waste ground. It has smaller flower-heads than the other species of *Lactuca* that grow in Britain, and they tend to close up around midday, making the plants harder to spot in the afternoon. Its leaves also lack the spines along the underside of the midrib seen in our other *Lactuca* species. It seeds are short-lived, making it vulnerable to harsh winters.

ALSO IN THE AREA

Saltmarsh goosefoot, *Chenopodium chenopodioides*, is largely restricted in Britain to the drier parts of salt-marshes around the Thames Estuary. See also pedunculate sea-purslane (*Atriplex pedunculatus*, p. 14).

CONSERVATION STATUS EN
FLOWERING Jul–Aug
PLANT HEIGHT 15–75 cm

Flower-heads of least lettuce. (CJD)

Old sea wall at Fobbing. (CJD)

Teesdale sandwort

Minuartia stricta

Teesdale sandwort is one of five British species in the genus *Minuartia*, all of them native and all of them rare. (Another rare species is native to western Ireland.) It is a perennial plant, forming small tufts up to 10 cm (4 in) tall.

Teesdale sandwort has a circumboreal distribution, with scattered occurrences further south, including in Britain and the Rocky Mountains. The species formerly occurred in the Alps and their foothills, and in the Jura mountains, but was thought to have died out there until a population was discovered in the Bavarian Alps in 2004. These more southerly populations are believed to be relics of a wider distribution following the last ice age.

The British population of Teesdale sandwort is restricted to Widdybank Fell in the limestone moorlands of upper Teesdale, at altitudes of 490–500 m (1,600–1,640 ft). The geology of the area includes outcrops of the rare 'sugar limestone', and Teesdale sandwort grows in wet flushes where they pass through this limestone. This area of occurrence was discovered in 1844, along with many other local rarities, although several populations were lost when the Cow Green Reservoir was created on the upper reaches of the River Tees in the 1960s for the benefit of the industries downstream. The scarcity of the habitat and the species' poor competitive ability may partly explain why it is now so rare. Although the site is designated as access land under the Countryside and Rights of Way Act 2000, that access has been temporarily suspended - until 2065 - to protect the fragile ecosystem.

Teesdale sandwort differs from spring sandwort (*Minuartia verna*), the only other species of *Minuartia* in the area, in that its flower-stalks are not hairy, and its leaves have only one faint vein. Some species of *Arenaria* may look similar, but have fruiting capsules that open by 6 or 10 teeth, rather than by 3 teeth as in Teesdale sandwort.

CONSERVATION STATUS	VU
FLOWERING	Jun–Jul
PLANT HEIGHT	3–10 cm

ALSO IN THE AREA

Upper Teesdale is home to several rarities, including spring gentian (*Gentiana verna*) and two microspecies of lady's-mantle (*Alchemilla monticola* and *Alchemilla subcrenata*) found nowhere else in Britain.

Teesdale sandwort. (CJD)

Wet flush through sugar limestone, Widdybank Fell. (CJD)

Holly-leaved naiad

Najas marina

Holly-leaved naiad is an annual plant of brackish water. It has a patchy distribution across much of the Northern Hemisphere, and is relatively common in Africa, where it occurs both in coastal regions and in rift lakes. It grows from roots in the muddy substrate, and produces unmistakable spiny-toothed leaves up to 5 cm (2 in) long, which give it its English name.

In 1883, Arthur Bennett, a specialist in water plants, found holly-leaved naiad growing in Hickling Broad, the largest of the Norfolk Broads. It now has apparently permanent populations in three broads, and transient populations appear each year in various others. Where it grows, it tends to form dense, tangled mats, with the stems branching and rooting at various points. In 2015, a colony inexplicably appeared at a Wildfowl and Wetlands Trust reserve near Arundel, West Sussex.

Holly-leaved naiad is a dioecious species, with separate male and female plants. It produces abundant seed in Britain, even though male plants weren't discovered until 1998 and even then the male flowers were mostly produced after the seeds had been set. It is possible that the majority of seeds are produced asexually.

For decades after its discovery, holly-leaved naiad grew rarer in Britain through the effects of pollution on the broads where it lived. This 'eutrophication' – loading with excess nutrients – may come from human sewage, run-off of agricultural fertilisers, or a combination of the two. Efforts are under way to curb the levels of nutrients, especially phosphorus, entering the Norfolk Broads, in order to improve the ecosystem for many plants and animals, not least the holly-leaved naiad.

ALSO IN THE AREA

The crested buckler-fern, *Dryopteris cristata*, is largely restricted in Britain to 'floating fens', where mildly acidic conditions develop on top of alkaline wetlands, mostly in the Broads.

CONSERVATION STATUS	VU
FLOWERING	Jul–Aug
PLANT HEIGHT	5–50 cm

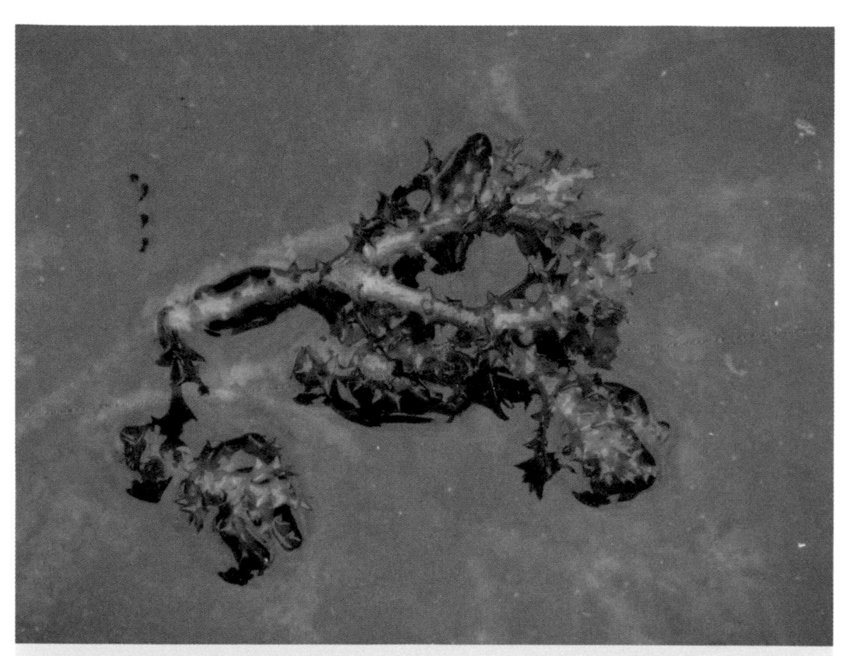

Holly-leaved naiad. (CJD)

Lakeshore at Hickling Broad. (CJD)

Least adder's-tongue

Ophioglossum lusitanicum

Least adder's-tongue is one of only four British species in the family Ophioglossaceae – the adder's-tongues and moonworts. This ancient group is allied to the ferns and horsetails, and has the highest chromosome counts of any organism (up to 1260 in the tropical species *Ophioglossum reticulatum*). The least adder's-tongue has around 250 chromosomes, far more than most plants and probably all animals. The reasons for this exceptional number of chromosomes are unclear. It is partly due to a number of polyploidisation events, in which the entire genome is doubled, that have occurred in the millions of years over which the group has existed – although there is normally strong selection pressure after such events to consolidate the chromosomes to a more manageable number.

Adder's-tongues have a simple above-ground structure: they produce a single leaf-blade, and another stalk that produces the spores (moonworts are similar, except that the leaf-blade is divided into paired lobes). Least adder's-tongue differs from the widespread common adder's-tongue (*Ophioglossum vulgatum*) and the small adder's-tongue (*Ophioglossum azoricum*) of western coastal areas by its small size and by appearing in winter or early spring, rather than in late spring and summer. The first factor may help to explain why it is so rarely found – few people would notice a plain green, flowerless plant less than 2 cm (0.8 in) high. Accordingly, it wasn't noticed in Britain until 1950, nearly a century after its first record from the Channel Islands.

Least adder's-tongue is fundamentally a warmth-loving species, albeit with a coastal bent, and Britain is at the northern limit of its range. Within Britain, it is only known from the Isles of Scilly, where it was first discovered by the classicist John Raven in 1950. Even within the Scillies, it only grows on the southernmost inhabited island of St. Agnes, in short, south-facing coastal turf.

CONSERVATION STATUS	VU
FLOWERING	Feb–Mar
PLANT HEIGHT	1–3 cm

ALSO IN THE AREA

The mild climate of the Scillies allows many plants to grow there that cannot survive in the rest of Britain. Most are aliens, but the orange bird's-foot, *Ornithopus pusillus*, is native. Others, such as the St. Martin's buttercup (*Ranunculus marginatus*) and western ramping-fumitory (*Fumaria occidentalis*) are also found on the Cornish mainland.

Least adder's-tongue growing in Calabria, Italy. (Domenico Puntillo; IC)

The top of Wingletang Down, St. Agnes. (Andrew Abbott; GG)

Military orchid

Orchis militaris

Military orchid closely resembles the monkey orchid (*Orchis simia*; see that species for differences). It is usually less than 45 cm (18 in) tall, and is named for the resemblance of the flowers to little soldiers, particularly because the five upper tepals together form a shape like an antiquated soldier's helmet.

Military orchid grows in calcareous grassland and woodland edges, especially over chalk, from northern Spain to Lake Baikal in eastern Russia. Changing agricultural practices were bad news for the military orchid and, by 1924, it had not been seen in Britain for around 10 years and was declared extinct.

It was therefore with some surprise that the botanist J.E. ('Ted') Lousley found a colony of around 40 plants at a lunch-stop in 1947, at a site that had recently been felled of trees as part of the war effort. Lousley kept the location a closely guarded secret, and when it was stumbled upon by the botanists Richard Fitter and Frances Rose in 1956, they sent Lousley a postcard stating only that '[t]he soldiers are at home in their fields'. (The site in question was later identified as Homefield Wood in the Chiltern Hills near Marlow, which is now protected as an SSSI and managed by the local wildlife trust.) There are also a couple of sites in the Chilterns of Oxfordshire where small numbers of military orchids occasionally appear.

Another remarkable discovery took place in 1955, when a colony of several hundred military orchids was found in Breckland, far outside the range where it had previously been known to grow. (That site is now the Rex Graham nature reserve, near Mildenhall, Suffolk.) The Chiltern populations were considered for a while to be a separate, exclusively British variety (*Orchis militaris* var. *tenuifrons*), but this distinction is no longer considered to be worth maintaining. The Suffolk population may nonetheless be a relatively recent arrival from continental Europe, independent of the other British plants. This would not necessarily mean that it had been introduced; the seeds of orchids are tiny, and can easily been transported over huge distances by natural means.

CONSERVATION STATUS	VU
FLOWERING	May–Jun
PLANT HEIGHT	25–45 cm

ALSO IN THE AREA

See under Suffolk lungwort (*Pulmonaria obscura*) for Breckland, and under fringed gentian (*Gentianopsis ciliata*) for the Chilterns.

Military orchid. (CJD)

Orchid meadow at Homefield Wood. (CJD)

Monkey orchid

Orchis simia

Monkey orchid is found from Europe and North Africa to Central Asia. Its flowers resemble little monkeys, with four limbs and a tail, and the whole plant is usually less than 30 cm (12 in) tall. The military orchid, *Orchis militaris* (q.v.), is similar but has lower 'limbs' that are thicker than the upper ones. A species where the bonnet covering the head is much darker than the limbs is known as the lady orchid, *Orchis purpurea*, and there is also a man orchid, *Orchis anthropophora* (formerly *Aceras anthropophorum*), which lacks the spur extending backwards from each flower. All of these can hybridise with each other in places where both parent species occur, producing variously intermediate offspring, which can make identification difficult.

In Britain, monkey orchid was first recorded from Park Gate Down in Kent in 1777. By 1955, no monkey orchids had been seen in Britain for over 30 years, and it was thought to have died out. In May that year, a single flowering spike was observed, but it was eaten by slugs before it could set any seed. Over the following years, more flowers appeared, but many were still chomped by slugs, rabbits and horses. Slug pellets and fencing helped to reduce the levels of herbivory, and manual pollination – using the stems of the upright brome (*Bromopsis erecta*) that grew nearby as a tool for transferring the pollen – led to an increase in the population. As well as the Kentish site, monkey orchid also grows at Hartslock nature reserve in Oxfordshire, where it occurs alongside the lady orchid, and hybridises with it. A further population persisted for a few years in south-eastern Yorkshire, but died out in 1983.

The decline in the monkey orchid's fortunes in Britain is mostly due to changing farming practices, such as the ploughing of the chalk downland where it generally grows. In some countries, including Turkey, it is much more common, and the fleshy tuber is dried and ground to make 'salep', a powder that can be used in cooking like arrowroot.

ALSO IN THE AREA
See under fringed gentian (*Gentianopsis ciliata*, p. 62).

CONSERVATION STATUS	VU
FLOWERING	May
PLANT HEIGHT	20–40 cm

Monkey orchid at Hartslock. (CJD)

Orchid meadow at Hartslock, with the River Thames behind. (CJD)

Bedstraw broomrape

Orobanche caryophyllacea

Bedstraw broomrape is a parasitic plant that grows on various hosts in the madder family (Rubiaceae), especially the hedge bedstraw (*Galium album*) and lady's bedstraw (*Galium verum*). It is a glandular-hairy plant up to 40 cm (16 in) tall, with deep red-purple stigmas, but pale flowers that smell of cloves, giving the species both its scientific name *caryophyllacea* and its alternative name of 'clove-scented broomrape'.

Bedstraw broomrape is found across much of Europe, as far east as the Caucasus and Iran. Within Britain, it is limited to a couple of areas of eastern Kent. The largest population grows on dunes around Sandwich Bay; other populations, consisting of up to 20 individuals each, grow in various kinds of scrubby ground near Folkestone. The fact that it is only found in one of the warmest corners of Britain suggests that its distribution is limited by temperature. Given that the species occurs in areas overseas where frosts are likely (such as Scandinavia and the foothills of the Alps), it is probably a need for hot summers rather than mild winters that keeps bedstraw broomrape restricted to Kent.

In the 18th century, bedstraw broomrape was mixed up with other species of broomrape under the now-rejected name '*Orobanche major*'. It took the English botanist James Smith, founder of the Linnean Society, to conclusively separate the bedstraw broomrape from the others. Ironically, he based his description on plants he had collected in central Italy, believing that bedstraw broomrape did not occur in Britain.

ALSO IN THE AREA

The late spider orchid (*Ophrys fuciflora*) is only found on the Kentish North Downs, close to Dover, and the early spider orchid (*Ophrys sphegodes*) has its British stronghold there, although it also occurs elsewhere in south-eastern England.

CONSERVATION STATUS	NT
FLOWERING	Jun–Jul
PLANT HEIGHT	20–40 cm

Bedstraw broomrape growing in Bavaria. (Bernd Haynold; WC)

Hedge bedstraw, *Galium album*, one of the main host plants for bedstraw broomrape. (CJD)

Oxtongue broomrape

Orobanche picridis

Oxtongue broomrape is an annual parasitic plant up to 60 cm (2 ft) tall that mainly attacks oxtongues (*Picris* and *Helminthotheca*). For a long time, the British plants were thought to belong in the species *Orobanche artemisiae-campestris*, but they are now considered part of a separate species, *Orobanche picridis*. (The narrow-sense *O. artemisiae-campestris* is a Mediterranean species.)

As a fully parasitic plant, oxtongue broomrape has no need for chlorophyll, and so produces only vestigial leaves with no green colouring. It can be distinguished from the other British species of broomrape by the shape of its sepals, which are all long and thin. The British populations are restricted to unstable coastal chalk cliffs, almost always in places where the hawkweed oxtongue, *Picris hieracioides*, grows; on the European mainland, it often grows away from the coast.

Within Britain, there are records of oxtongue broomrape from a number of areas (including the site of its first discovery, by Rev. W. W. Newbould in Cambridgeshire in 1878), but only two areas seems to still have populations. The first is in Kent, along part of the white cliffs from Dover, through St. Margarets at Cliffe, to Kingsdown, where the cliffs finally subside and are replaced by shingle beaches. The second population lives on the sea-facing chalk downs near Freshwater on the Isle of Wight. A further population of broomrapes on the floodplain of the River Adur in Sussex was photographed in 1986, but could not be re-found after it had been identified to the species level.

Because of the instability of the cliffs, including frequent cliff-falls, the plants are at constant risk of uprooting, and it is difficult to know how many plants actually exist in any given year. The sizeable length of unstable chalk cliffs along the south coast of England does, though, raise the possibility that further populations have gone undetected.

CONSERVATION STATUS	EN
FLOWERING	Jun–Jul
PLANT HEIGHT	10–50 cm

Oxtongue broomrape growing on the Isle of Wight. (CJD)

Hawkweed oxtongue, *Picris hieracioides*, the main host plant for oxtongue broomrape. (CJD)

Yellow oxytropis

Oxytropis campestris

Yellow oxytropis is a low-growing perennial legume mostly found in well-drained mountainous areas on fairly bare calcareous rock. It produces erect groups of flowers, each flower being 15–20 mm (0.6–0.8 in) long. The name 'oxytropis' is Greek for 'pointed keel', and refers to the shape of the normally concealed 'keel' petal in each flower. The genus *Oxytropis* contains more than 300 species worldwide, but only two occur naturally in Britain, both of them rare and restricted to Scotland. The most obvious difference between the yellow oxytropis and the purple oxytropis is the flower colour, although the yellow oxytropis sometimes produces purple flowers and *vice versa*. (The most reliable difference between the two concerns the orientation of the membranes within the seed-pod.)

Yellow oxytropis is an arctic-alpine plant from around the Northern Hemisphere. Within Europe, it is found in Scotland and parts of Scandinavia and, quite separately, in the more southerly mountains from the Pyrenees to the Caucasus. Within Scotland, it has a curious distribution. Two sites occur at relatively high altitude (500–650 m; 1,600–2,100 ft) in the central Highlands – one in Glen Doll discovered in 1812 by George Don (father of the botanists George and David Don) and one on cliffs above the imaginatively named Loch Loch discovered by the Rev. J. Fergusson in 1887 – while a third, previously misidentified as purple oxytropis, occurs below 180 m (590 ft) on sea-cliffs in Kintyre.

Species of *Oxytropis* and its close relative *Astragalus* are known as 'locoweed' in North America, in reference to their toxic effects on grazing animals. This toxicity is a significant problem in North America, but may also prevent yellow oxytropis from being grazed by goats in Scotland. The two upland Scottish populations at least are relatively large and stable, despite being extensively collected by amateur botanists.

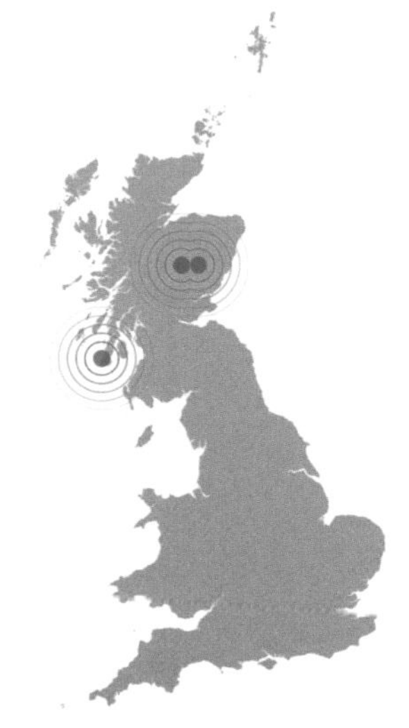

CONSERVATION STATUS	VU
FLOWERING	Jun–Aug
PLANT HEIGHT	10–20 cm

ALSO IN THE AREA

Alpine milk-vetch, *Astragalus alpinus*, is a similar plant to yellow oxytropis, but with purple-tipped flowers, and grows in similar areas in the eastern Highlands.

Yellow oxytropis growing in the Pyrenees. (Réginald Hulhoven; WC)

The cliffs of Corrie Fee. (CJD)

Childing pink

Petrorhagia nanteuilii

Childing pink is one of three annual species of *Petrorhagia* that occur in Britain, and the differences between the three are slight. The proliferous pink, *Petrorhagia prolifera*, may be native in eastern England, whereas *Petrorhagia dubia* has been recently introduced to Hampshire, and is now known as Hayling Island pink. The best way to distinguish the childing pink from the Hayling Island pink is to examine the seeds under a microscope; the childing pink's seeds are covered in small bumps, whereas the Hayling Island pink's seeds have a network of ridges. It is perhaps understandable that the two species have been mixed up in the past, making it hard to pinpoint the first record of any of the species. Complicating matters further, childing pink is thought to be an allopolyploid hybrid of the Hayling Island pink and the proliferous pink.

All three annual species of *Petrorhagia* are thin, erect plants with straight, narrow, rather bluish leaves and a flower-head surrounded by papery bracts. Childing pink is typically 10–20 cm (4–8 in) high, and pretty pink flowers emerge from within the bracts in June and July, normally one at a time. Its seeds are large, and drop close to the parent plant, limiting the species' ability to spread to new sites.

Childing pink is a plant of the western Mediterranean, chiefly the Iberian Peninsula, but extending along the coastlines to Jersey, North Africa and western Asia. In Britain, it is only found naturally at Shoreham-by-Sea and at Pagham Harbour, between Bognor Regis and Selsey Bill in West Sussex. (An introduced population has persisted in South Wales since the 1930s.) At Pagham, it grows on sand and shingle bars, one of which is closed to visitors in the summer to protect breeding birds. The greatest threat to the childing pink at Pagham may come from another Mediterranean plant, red valerian (*Centranthus ruber*), which outcompetes many plants, including the childing pink; at the Shoreham site, tall grasses

CONSERVATION STATUS	VU
FLOWERING	Jun–Aug
PLANT HEIGHT	20–50 cm

and three-cornered garlic (*Allium triquetrum*) could also suppress the childing pink, and are being removed.

ALSO IN THE AREA

The red star-thistle, *Centaurea calcitrapa* – a kind of very spiny knapweed – is sometimes thought to be native to the coasts of Sussex, although it is usually considered an alien.

Childing pink growing on Minorca. (Liliane Roubaudi; TB)

Vegetated shingle at Pagham Harbour. (CJD)

Blue heath

Phyllodoce caerulea

Blue heath is a dwarf shrub in the heather family, Ericaceae, with flowers 7–12 mm (0.3–0.5 in) long. Although it is called the blue heath and is often depicted with blue flowers, they are better described as purple. It is found at high northern latitudes around the globe, extending as far south as parts of Japan. In Europe, it grows in much of Scandinavia, with scattered outposts in Scotland and the Pyrenees; a reported occurrence in the Swiss Alps may be an error. Blue heath resembles many other species in the heather family, especially bog rosemary, *Andromeda polifolia*. In contrast to bog rosemary, though, blue heath has gland-tipped hairs on its flower stalks and sepals.

For many years, blue heath was thought to be restricted within Britain to the mountain known as the Sow of Atholl (more properly Meall an Dobharchain), an 800-m (2,600-ft) rounded lump beside the Pass of Drumochter. There were also early, unsubstantiated reports from near Aviemore and Fort William and, less plausibly, the tiny Shiant Isles in the Outer Hebrides. The plant's numbers have varied over the decades, and it has been (wrongly) declared extinct in Britain a number of times. From 1970, a few further populations were discovered a short distance west of the Sow of Atholl, around the Ben Alder Forest.

The populations in Scotland are at risk from climate change. Blue heath requires winter snow cover in order to flower and set seed in the wild (although it thrives in cultivation in southern England). The known populations mostly grow in areas that receive a lot of snow, in sites sloping north or north-east such that they hardly receive any direct sunlight except at the height of summer. Their combined population is probably around 300. The largest population is on the Sow of Atholl, but here the plants are so scattered that they are revealed by the melting snow in summer at different times, and so flower at different times, preventing cross-pollination.

CONSERVATION STATUS	VU
FLOWERING	Jun–Jul
PLANT HEIGHT	5–25 cm

ALSO IN THE AREA

Beyond the species covered elsewhere in this book, the wider area around Ben Alder also contains two rare sedges – the mountain bog-sedge (*Carex rariflora*) and the scorched Alpine sedge (*Carex atrofusca*).

Blue heath growing in Lapland. (Alinja; WC)

The Sow of Atholl seen from the Pass of Drumochter. (Mick Knapton; WC)

American pondweed

Potamogeton epihydrus

Within Europe, American pondweed is only considered native to Britain, although it occurs across much of North America. In the UK, it has been introduced to canals on either side of the Pennines, but only occurs natively in a few small, shallow lakes near the middle of South Uist in the Outer Hebrides. These populations are genetically uniform, suggesting that they are the product of very few founding individuals.

There are many species of pondweed in Britain, and distinguishing them is not always easy. American pondweed produces both floating leaves (up to 8 cm or 3 in long) and submerged leaves, and the submerged leaves are very long and narrow, like blades of grass.

American pondweed was first discovered in Britain in 1907 when a Miss A. Vigiers found it growing in the Calder and Hebble Navigation at Halifax, West Yorkshire. It now occurs throughout the stretch from Sowerby Bridge to Halifax. It also occurs in the Rochdale Canal near Oldham; in both cases, it is thought to have been introduced unwittingly from America, probably with cotton, which was imported on a huge scale to the mills of northern England at the time. The Uist population was first discovered in 1943, by J. W. Heslop-Harrison. A 1951 radio broadcast that seemed to suggest that it had been first found that year seems to have incensed Heslop-Harrison, who reiterated his earlier discovery. He did this initially in *The Vasculum*, a local journal of which he was then the editor, and followed it up with a more sober paragraph in *Nature*.

It is not clear how American pondweed arrived in North Uist; Heslop-Harrison stated that '[s]ince the favoured habitats in South Uist are to be found far from human habitation of any sort on desolate moorlands, any suspicion of introduction ... is untenable'. Normally, that would be the end of the matter, and speculation would turn to the transport of seeds on the feet of waterfowl, but Heslop-Harrison had form.

Heslop-Harrison was a great botanist, but doubt has been cast on several of his more sensational

CONSERVATION STATUS	VU
FLOWERING	Jul–Aug
PLANT HEIGHT	10–90 cm

finds from the Hebrides, especially those on the Isle of Rùm. It was suspected that plants were being brought in from abroad and planted, either by Heslop-Harrison or by pranksters among his students. Several of his reports have been discredited through a lack of evidence. His successes may also have been uncomfortable to a scientific elite who looked down on a working-class man with ideas above his station (the double-barrelled surname was his own invention, for instance). In the case of the American pondweed, there can be no doubt that it does indeed grow on South Uist, and that Heslop-Harrison found it first. We may never know for certain how it got there.

American pondweed growing in New Hampshire. (Barre Hellquist; EM)

Shallow moorland lakes on South Uist. (Rob Burke; GG)

Suffolk lungwort

Pulmonaria obscura

Suffolk lungwort, also known as unspotted lungwort, is a perennial herb in the forget-me-not family, Boraginaceae, with flowers around 15 mm (0.6 in) wide. Most lungworts have white spots on their leaves, but Suffolk lungwort has only faint green spots, if any; it differs from the other unspotted lungworts in Britain in that its lowest leaves are heart-shaped (cordate) at the base. The history of Suffolk lungwort in Britain has been clouded by confusion with other species and by the misapplication of names. It is most frequently confused with the common lungwort, *Pulmonaria officinalis*, which is a common garden plant and a widespread alien in the wild.

Suffolk lungwort is widely distributed across central and eastern Europe, especially in areas with a relatively continental climate. In Britain, it only occurs in the heart of East Anglia, the area with the most continental climate, where it was first recognised in 1842. These occurrences are most likely to be natural, rather than introduced, because they occur in natural situations and the species is not widely grown in cultivation.

Three populations are known, all in a small area of Suffolk, and all in remnant patches of ancient woodland on private land. A fourth population once existed at Millfield Wood near Polstead Heath, 20 miles (32 km) west of the others, but that patch of woodland was largely destroyed to make way for electricity pylons. The largest of the populations was estimated to have produced 86 flowering stems in 1979, a number that probably closely reflects the current situation. The main conservation measure in place is appropriate woodland management. Suffolk lungwort seems to be most abundant in more recently coppiced parts of the woods where it lives, which suggests that continued coppicing may be necessary to ensure its long-term survival in Suffolk.

ALSO IN THE AREA

Breckland rarities include the perennial knawel (*Scleranthus perennis*, p. 112), Breckland thyme (*Thymus serpyllus*), spring speedwell (*Veronica verna*, p. 132) and two other non-native

CONSERVATION STATUS	EN
FLOWERING	May–Jun
PLANT HEIGHT	10–30 cm

speedwells (Breckland speedwell, *Veronica praecox*, and fingered speedwell, *Veronica triphyllos*), Spanish catchfly (*Silene otites*) and the purple-stemmed cat's-tail (*Phleum phleoides*).

Suffolk lungwort growing in Germany. (CJD)

Ancient woodland near Burgate, Suffolk. (CJD)

Adder's-tongue spearwort

Ranunculus ophioglossifolius

Adder's-tongue spearwort is a species of annual buttercup that grows in marshy conditions. Its distribution is chiefly Mediterranean, but it also grows in numerous scattered locations that become scarcer to the north. In Scandinavia, for instance, it only occurs on the Swedish island of Gotland, and in Britain, it was discovered by the stonewort specialist Henry Groves in Hampshire in 1878 but is now only found at two sites in Gloucestershire.

Adder's-tongue spearwort has the typical yellow flowers of a buttercup or spearwort, albeit fairly small (5–9 mm or 0.2–0.4 in across), and leaves that are heart-shaped at the base and toothed, but not divided into distinct lobes. It grows upright, unlike the closely related lesser spearwort, *Ranunculus flammula*.

Adder's-tongue spearwort is quite exacting in its ecological demands: it grows in places that are inundated but frost-free (at least below the water surface) in winter, are bare and damp in spring, and have lower water levels in summer. It used to occur at sites in Dorset and Hampshire, but it has not been seen at either for a long time and has become reduced in Britain to two sites in Gloucestershire, with the larger population at Badgeworth nature reserve, near Cheltenham. Conservation measures were begun quite early at Badgeworth and the site has been managed since 1962 specifically to enable adder's-tongue spearwort to thrive; the reserve is open to the public for a few hours on one day a year. The second population, in the Lower Woods near Wickwar, is subject to grazing and less intense conservation, and numbers there are less stable.

The seeds of adder's-tongue spearwort can remain viable for up to 30 years, allowing the population to persist over one or more bad years. In a good year, there may be as many as a thousand flowering plants at Badgeworth; in poor years, none.

ALSO IN THE AREA

Downy-fruited sedge, *Carex filiformis*, grows in the Cotswolds and at a few other sites in southern England.

CONSERVATION STATUS	VU
FLOWERING	Jun–Jul
PLANT HEIGHT	15–40 cm

Adder's-tongue spearwort. (CJD)

Badgeworth nature reserve. (CJD)

Sand crocus

Romulea columnae

Sand crocus is a perennial plant, regrowing from a corm every autumn. It grows from Turkey to the Azores, and along the Atlantic coast as far as south-western Britain. It is a low-growing species of closely grazed coastal swards, producing long overwintering leaves in autumn and flowering in spring. The flowers range from pale purple to white, becoming yellow towards the centre of the flower. The whole flower is less than an inch across, and often less than half that.

Sand crocus differs from the genus *Crocus* by its thick, unstriped leaves. Although the sand crocus grows from corms, it reproduces mainly through seed, rather than by division of the corm. It needs open vegetation to allow the seeds to germinate and the young plants to receive enough sunlight.

Sand crocus is relatively common in the Channel Islands, but in Britain it is restricted to two sites in the south-west. It grew on cliff-tops at Polruan in Cornwall in the late 19th century, but appeared to have died out until it was rediscovered in 2002, after a gap in records of more than 120 years. When rediscovered, the population was estimated at around 1,500 plants. In Devon, it is restricted to a golf course at Dawlish Warren and is also known as the 'warren crocus'. It was discovered there in 1834, and is present as a number of scattered colonies. One such colony was in an area destined to be made into a car park, but the turf was taken up and transplanted elsewhere, and that colony continues to thrive in its new location.

CONSERVATION STATUS	VU
FLOWERING	Apr–May
PLANT HEIGHT	1–3 cm

Sand crocus growing in the French Riviera. (Errol Vela; TB)

Dune vegetation at Dawlish Warren. (Des Blenkinsopp; GG)

Scottish dock

Rumex aquaticus

Scottish dock is a perennial herb in the dock family, Polygonaceae, growing up to 2 m (6.5 ft) high. Its leaves are broad at the base, and heart-shaped where they meet their stalks, but it can be best told apart from its relatives by the form of its fruits. These are longer than they are wide, without long teeth and with no tubercles (most other docks have a conspicuous lump, or tubercle, on one or more of the fruit's three sides).

Scottish dock is found from northern and central Europe across to Central Asia; a related species, *Rumex occidentalis*, continues that range around to eastern North America. Within Britain, Scottish dock is restricted to the area around Loch Lomond, where it was first discovered in 1935. It grows in shallow, seasonally inundated areas, generally close to the mouths of small rivers. The largest populations are around the entrance of Endrick Water into the lake, but others are near the mouths of Finglas Water and Fruin Water. A further population is known from further up the Endrick Water, near the hamlet of Croftamie.

Part of the reason this species was overlooked in Britain for so long may be because the name '*Rumex aquaticus*' had been misapplied to other species of dock for many years, including water dock, a much more frequent plant that is perhaps more deserving of the epithet '*aquaticus*' but is correctly called *Rumex hydrolapathum*. Although ecologically similar to the Scottish dock, the leaves and fruits of water dock set it apart.

Broad-leaved dock, *Rumex obtusifolius*, prefers drier locations than Scottish dock, but is a close enough relative that the two will hybridise. Because the broad-leaved dock is far more abundant, there are concerns that repeated hybridisation could swamp the Scottish dock with genes from the commoner species, and drive it effectively to extinction. It is hoped that by maintaining the wetter habitats favoured by Scottish dock, it will be able to fend off the broad-leaved dock in the longer term.

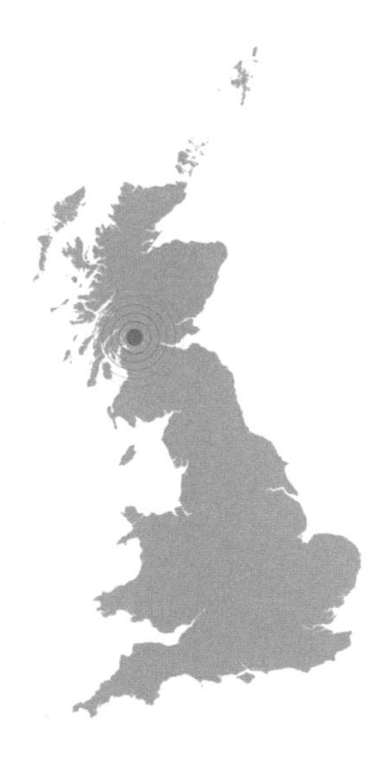

CONSERVATION STATUS	VU
FLOWERING	Jul–Aug
PLANT HEIGHT	100–200 cm

ALSO IN THE AREA

Within Britain, tufted loosestrife, *Lysimachia thyrsiflora*, is mostly found in wetlands in the Scottish Central Belt, especially along canals.

Scottish dock. (CJD)

Marsh close to Endrick Water. (CJD)

Drooping saxifrage

Saxifraga cernua

Drooping saxifrage is a distinctive species of perennial herb up to 15 cm (6 in) tall that grows in calcareous mountains. It is found almost throughout the Arctic, and is rare and patchily distributed in the Alps, Carpathians and Rocky Mountains.

In Britain, drooping saxifrage is only found on some of the highest mountains, at altitudes above 830 m (2,700 ft). (Only the hare's-foot sedge, *Carex lachenalii*, might have a lower limit at greater altitude in Britain.) It was first recorded by James Dickson on Ben Lawers in 1790, and is now known from three quite distant areas – Ben Lawers, Glen Coe and Ben Nevis / Aonach Beag. In each case, the drooping saxifrage tends to occur in sites below the most exposed ridgelines, where snow lies late into the year.

Drooping saxifrage is unique among British saxifrages in that almost all of its flowers are replaced by bulbils, specialised growths that can drop off and form new plants. There may be a single flower up to an inch (25 mm) across at the tip of each stem, but these rarely produce pollen in Britain, and so few seeds are produced, meaning that reproduction is almost entirely achieved through the bulbils. Each of the Scottish populations is therefore thought to represent a single clone, although there may be nearly 2,000 plants across the three areas. They are not thought to be in great direct danger, although occasional grazing and trampling may destroy a few plants at a time. The greatest threat is probably climate change; with little chance of migrating to higher altitudes, drooping saxifrage may suffer if the temperature rises significantly.

ALSO IN THE AREA

In the middle of the range of the drooping saxifrage is Rannoch Moor, the British stronghold for the Rannoch rush, *Scheuchzeria palustris*.

CONSERVATION STATUS	VU
FLOWERING	Jul–Aug
PLANT HEIGHT	10–20 cm

Drooping saxifrage growing in Greenland. (Kim Hansen; WC)

Bidean nam Bian, Glencoe. (Trevor Littlewood; GG)

Triangular club-rush

Schoenoplectus triqueter

Triangular club-rush is a species of tall herb in the sedge family that grows in brackish water. It is one of two club-rushes in Britain with triangular stems (a common feature in the family), the other being a rare exotic species not found in the same vicinity as this species. Triangular club-rush can reach 1.5 m (5 ft) tall, generally without lateral leaves, and with a group of inflorescences sprouting from the side of the stem, near the top. It can hybridise with other, more common species of club-rush, and the hybrids can approach either parent species in their appearance.

Triangular club-rush was first found in Britain in the 1650s at the old Horse Ferry between Westminster and Lambeth (replaced in 1862 by the first Lambeth Bridge), and has been recorded over much of the tidal stretch of the Thames, from Richmond-upon-Thames to Greenwich. Gradually, though, with the increasing economic development of London, and especially the building of embankments along the Thames, the populations died out, the last of them in 1946. Populations on the River Medway in Kent and the River Arun in West Sussex have likewise disappeared, and the species is now found in Britain only on the banks of the River Tamar on the Devon-Cornwall border. (It also occurs on the River Shannon in western Ireland.) Here, it is being outcompeted by the common reed, *Phragmites australis*, and it may have been reduced to two clonal clumps. For some reason, wild plants in Britain produce few viable seeds, although plants in cultivation seed as profusely as those growing overseas.

ALSO IN THE AREA

The lesser tongue orchid, *Serapias parviflora*, was found in 1989 at Rame Head, close to the Tamar estuary. It remains unclear whether it arrived naturally or through human interference.

CONSERVATION STATUS CR
FLOWERING Jul–Sep
PLANT HEIGHT 50–150 cm

Triangular club-rush. (Krzysztof Ziarnek; WC)

River Tamar at Morwellham. (CJD)

Brown bog-rush

Schoenus ferrugineus

Brown bog-rush is one of two species in the genus *Schoenus* in Britain. It differs from the black bog-rush, *Schoenus nigricans*, in having shorter leaves and fewer flower-heads, with the bract at their base being little longer than the inflorescence; in the black bog-rush, the bract is much longer.

Brown bog-rush is a perennial plant that usually forms tussocks up to 40 cm (16 in) high in wet flushes running through calcareous grassland. It expands vegetatively, and also produces seeds, but these rarely travel far; they germinate while still on the plant, when the stem eventually withers to the ground. For some reason, Scottish plants produce fewer viable seeds than plants growing in Sweden.

Brown bog-rush has a scattered distribution from Scandinavia to the mountains of southern Europe. In Britain, it is restricted to six low-lying sites in central Scotland. It was discovered at Loch Tummel in 1884, which remained its only known location for a long time. In 1950, the Clunie Dam was built at the loch's eastern end, raising the water level by 4.5 m (nearly 15 ft) and wiping out the population of brown bog-rush. Fortunately, some plants had been moved to a site on Ben Vrackie, and a number of additional populations have since been discovered. The Scottish population of brown bog-rush now stands at around 12,000, a large majority of which are at a single site. The species' unnoticed survival at so many sites suggests that there may be other, unknown populations in the area.

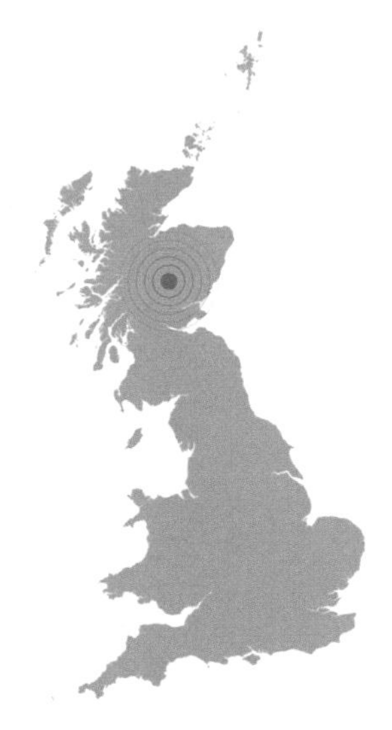

CONSERVATION STATUS	LC
FLOWERING	Jun–Jul
PLANT HEIGHT	15–30 cm

Brown bog-rush growing in Provence, France. (Liliane Roubaudi; TB)

Ben Vrackie from the south-west. (Gortyna; WC)

Round-headed club-rush

Scirpoides holoschoenus

Round-headed club-rush is a perennial herb in the sedge family, Cyperaceae, that lives in dune-slacks in Britain and also in other open habitats overseas. It is found from the Canary Islands to Greece, and as far north as Britain and central Russia.

Round-headed club-rush grows up to 1.5 m (5 ft) tall, with green stems and few leaves. Its flowers are grouped into spikelets, as in the rest of the family, but the spikelets are then aggregated into tight, spherical clusters. These are unique within the family, and only approached in appearance by the quite different bur-reeds of the genus *Sparganium*.

Within Britain, round-headed club-rush has been found as an alien plant at several sites in southern England and South Wales (including one persistent clump in Poole Harbour), but is only native at two sites in the south-west. It has been known since 1688 at Berrow Dunes in Somerset, where a single clump, 4 m by 5 m (13 ft by 16 ft), has survived despite a number of assaults, including its surroundings being converted into a golf course. Its stronghold, though, is at Braunton Burrows on the North Devon coast, where there are more than 3,000 clumps.

Round-headed club-rush flowers in August and September, and only sets seed after long, hot summers, which may explain the species' limited distribution in Britain.

ALSO IN THE AREA

Berrow also contains a population of the Somerset rush, *Juncus subulatus*, first recorded in 1957. No one knows how the species arrived in Britain from the Mediterranean region, or therefore whether it is native or introduced here. Braunton Burrows are also home to the largest British population of the water germander, *Teucrium scordium* (p. 126).

CONSERVATION STATUS	EN
FLOWERING	Aug–Sep
PLANT HEIGHT	30–150 cm

Round-headed club-rush. (CJD)

Habitat in Berrow Dunes. (CJD)

Perennial knawel

Scleranthus perennis

A number of British plants can be divided into a widespread subspecies and a rare, local subspecies. Perennial knawel is unique among the rare plants of Britain in that it is made up of two subspecies, both of which are extremely rare here. It is a spreading perennial plant with thin, pointed leaves, growing in bare areas and beside paths and tracks. Its flowers have no petals, but the five green sepals each have a strong white border. The only similar plant in Britain is its close relative the annual knawel (*Scleranthus annuus*), which has more sharply pointed sepals with a narrower white border, and only lives for one year.

One subspecies, *Scleranthus perennis* subsp. *perennis*, is widespread across most of continental Europe and can grow up to 20 cm (8 in) high. In Britain, it is only found at Stanner Rocks in Radnorshire, where it grows in alkaline conditions. Perennial knawel is normally a short-lived plant, relying on its seed bank for its long-term survival. This strategy can be risky, and the population size at Stanner Rocks varies dramatically from year to year, and probably never exceeds 150.

The second subspecies, *Scleranthus perennis* subsp. *prostratus*, grows on acid, sandy soils in Breckland. It was first found there by John Ray in 1696, and occurs nowhere else in the world. It is much shorter than the other subspecies, and has smaller seeds. It has dwindled in Breckland to a handful of populations, although these include one at RAF Lakenheath of several thousand plants. Some former sites have been lost to housing development, intensive agriculture or forestry, but it has also been introduced to additional sites within its native range, with some success.

ALSO IN THE AREA

See Suffolk lungwort (*Pulmonaria obscura*, p. 96) for Breckland, and Radnor lily (*Gagea bohemica*, p. 56) for Stanner Rocks.

CONSERVATION STATUS	CR/EN
FLOWERING	Jun–Aug
PLANT HEIGHT	5–20 cm

Scleranthus perennis subsp. *perennis* growing in the Maritime Alps. (CJD)

Track through Breckland. (Hugh Venables; GG)

Viper's-grass

Scorzonera humilis

Viper's-grass is a plant in the group known as 'damned yellow composites' – members of the tribe Cichorieae in the family Compositae (or Asteraceae; both names are valid) that have yellow inflorescences and can be difficult to tell apart. Among the British members, however, viper's-grass is the only one to have a downy covering of soft hairs on the young stems. Its leaves are untoothed, which is uncommon in the group, and its phyllaries (the leaf-like bracts that surround the base of each flower-head) get progressively shorter towards the outside.

It flowers from May to the end of June, with flower-heads 2–3 cm (0.8–1.2 in) across, but dies back quickly, so that by August there may be little sign of the plant above the ground. Below ground, though, a thick black taproot persists over the winter and resprouts in spring.

Viper's-grass is widespread but increasingly rare across much of Europe, becoming confined to higher ground towards the south. Within Britain, it is only found at two sites in Glamorgan and one in Dorset, always in damp grassland or fen-meadows. Another population in Warwickshire persisted for a few years in the 1950s and 1960s. The Dorset site was the first to be discovered, in 1914, and is now part of the RSPB's Wareham Meadows reserve; a few lonely individuals have occasionally been found at other sites nearby. The Glamorgan sites were only discovered in 1996 and 1997, although the population at Cefn Cribwr Meadows covers several hectares and includes thousands of plants. The exact location of the second Welsh site, in the north-eastern part of the Gower Peninsula, has not been made public. The rarity of the nutrient-poor lowland grassland habitat may partly explain the rarity of viper's-grass, but other factors may also be involved. A scarcity of pollen may limit the seed set, and inbreeding will progressively reduce the plant's fitness, especially in small populations.

CONSERVATION STATUS VU
FLOWERING May–Jul
PLANT HEIGHT 10–40 cm

ALSO IN THE AREA

The fen orchid, *Liparis loeselii*, is a rare orchid with greenish-yellow flowers that lives in Glamorgan, Devon and Norfolk.

Inflorescence and leaves. (CJD)

Damp meadows at Cefn Cribwr. (CJD)

Cambridge milk-parsley

Selinum carvifolia

Cambridge milk-parsley is a member of the parsley and carrot family, most of which have very similar groups of flowers, radiating from a central stalk. Fruit are needed in order to make a firm identification, and those of Cambridge milk-parsley are hairless and flattened, 3–4 mm long, with tall ridges running along the centre, similar to those of *Angelica* and several other genera. Unlike Cambridge milk-parsley, *Thyselium palustre* (milk-parsley) has hollow stems, and *Angelica* has much larger leaf-lobes than Cambridge milk-parsley (which is itself known in North America as 'little-leaf angelica').

Cambridge milk-parsley grows in fens – alkaline marshes – and has been found at six sites in eastern England, although it is now restricted to three, all in Cambridgeshire. These populations are thriving under careful management, and they each hold several hundred or thousand individual plants; exact numbers are hard to come by, because the plants only start to flower two or three years after they germinate. The flowers appear from July to September, and the fruits are produced in October.

Cambridge milk-parsley was first found in Britain in 1880, when the Rev. W. Fowler discovered it growing near Broughton, Lincolnshire. He predicted that it would be found elsewhere in eastern England, and was soon vindicated when it was found in Cambridgeshire in 1882. It is odd that the plant, up to a metre (3 ft) tall, should have been overlooked by the botanists of Cambridge for so long, and it has been suggested that none of them had even visited Chippenham Fen until *Selinum carvifolia* was found there. A correspondence on the subject suggests that the local luminaries petulantly wrote it off as an introduction, rather than admit their own oversight. George Claridge Druce – a botanist from Oxford University – considered it to be native and noted that its close resemblance to the wild carrot, *Daucus carota*, may have helped it pass undetected; he wrote charitably that 'even careful and keen-eyed men may have overlooked it'.

CONSERVATION STATUS	VU
FLOWERING	Jul–Aug
PLANT HEIGHT	30–100 cm

ALSO IN THE AREA

The rare plants of Fenland include fen ragwort (*Senecio paludosus*, p. 118), water germander (*Teucrium scordium*, p. 126), fen violet (*Viola persicifolia*, p. 136), a subspecies of the heath violet (*Viola canina* subsp. *montana*), and the possibly extinct fen woodrush (*Luzula pallidula*).

Cambridge milk-parsley flowering and fruiting, in cultivation. (Andrea Moro; IC)

Fen vegetation in Cambridgeshire. (CJD)

Fen ragwort

Senecio paludosus

Senecio is the largest genus in the largest family of flowering plants, and contains over 1,000 species, although it is in the process of being carved into more manageable chunks. Some of these species are very common – arguably, too common – such as the native common ragwort (*Senecio jacobaea*, or *Jacobaea vulgaris* when *Senecio* is divided up) or the invasive Oxford ragwort, *Senecio squalidus*. The Oxford ragwort arose in the University of Oxford Botanic Garden by hybridisation between two species from Sicily, and has since spread across most of Britain, and hybridised with native species to produce the endemic hybrids *Senecio cambrensis* (Welsh groundsel or Welsh ragwort) and the now-extinct *Senecio eboracensis* (York groundsel).

The fen ragwort (*Senecio paludosus*, or *Jacobaea paludosus* when *Senecio* is split) is quite distinctive within this difficult genus, with its long, narrow leaves with sawtooth edges and entirely green rather than black-tipped phyllaries (bracts around the base of the flower-head). Its flower-heads have yellow rays 10–15 mm (0.4–0.6 in) long. The hairiness of the underside of the leaves also helps to separate this from the Oxford ragwort, which has almost hairless leaves. Fen ragwort grows in fens and reed-swamps, where it depends on regular clearing to survive.

Fen ragwort is found across much of central and eastern Europe, extending into central France, with much more scattered occurrences to the west and north. It was one of a number of species first discovered in Britain by John Ray, who found it in 1660. Its range gradually decreased until it vanished from its last site in the middle of the 19th century. For more than 100 years, it was thought to have died out in Britain, until a small population with only three flowering stems was found growing in a roadside ditch near Ely (Cambridgeshire) in 1972. The precise location is kept secret, partly because its position beside a road makes it easily accessible and therefore vulnerable. Curiously, the ditch had only been dug in 1968, suggesting that the seed may be viable for a long time, and that the species could reappear elsewhere. Seed

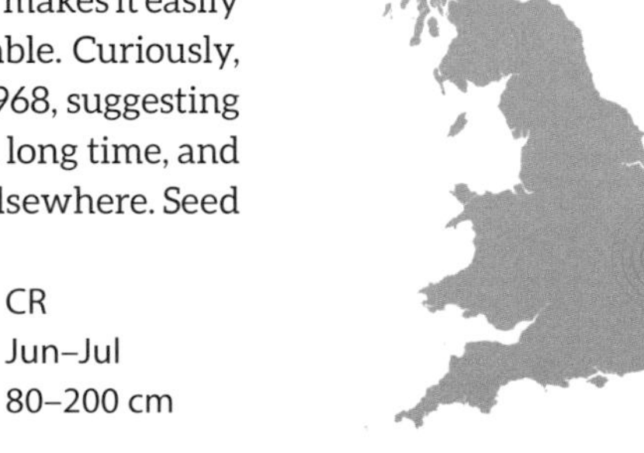

CONSERVATION STATUS	CR
FLOWERING	Jun–Jul
PLANT HEIGHT	80–200 cm

has been collected and cultivated at nearby sites, including Wicken Fen and Woodwalton Fen.

ALSO IN THE AREA

See under Cambridge milk-parsley (*Selinum carvifolia*, p. 116).

Fen ragwort growing in eastern Europe. (Krzysztof Ziarnek; WC)

Damp roadside verge near Ely. (CJD)

Moon carrot

Seseli libanotis

Moon carrot is another umbellifer, a member of the carrot and parsley family. Most genera in the family contain few species, but *Seseli* is one of the largest, with more than 100 species across the world.

Moon carrot is a biennial plant, or a short-lived perennial that flowers once before dying, that grows up to 60 cm (2 ft) tall. It closely resembles the wild carrot, *Daucus carota*; its fruits are 2.5–3.5 mm long and have much finer hairs than those of the carrot when examined closely. In late July, however, moon carrot can sometimes be identified at a distance by the faintly green colouration of its inflorescences; those of wild carrot and burnet saxifrage (*Pimpinella saxifraga*) are a purer white overall.

Moon carrot grows in chalk grassland across temperate Eurasia, from Spain to Japan. There is significant variation between populations across the species' range, particularly in the plant's hairiness, but the various subspecies that have been described are not widely recognised. Moon carrot is restricted in Britain to six sites in three separate areas of south-eastern England. The greatest concentration of plants occurs at Knocking Hoe nature reserve in Bedfordshire, where up to 12,000 individuals may be found, and at adjacent sites extending into Hertfordshire. Two populations of around 3,000 each live at Beachy Head, East Sussex, and several much smaller populations occur in Cambridgeshire, at Cherry Hinton and on the Gog Magog Hills. This last site is where the species was first found in Britain, by the Cambridge-based clergyman–botanist John Ray in 1690. All the populations are in protected areas, and they seem to be generally in good health.

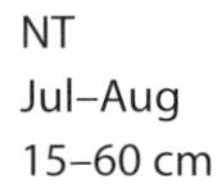

CONSERVATION STATUS	NT
FLOWERING	Jul–Aug
PLANT HEIGHT	15–60 cm

ALSO IN THE AREA

Meikle Kilrannoch is also home to the entire population of an endemic subspecies of the common mouse-ear, *Cerastium fontanum* subsp. *scoticum*.

Alpine catchfly. (Réginald Hulhoven; WC)

Hobcarton Crag. (CJD)

Downy woundwort

Stachys germanica

Downy woundwort is a biennial plant in the mint and sage family, growing up to 80 cm (32 in) tall. It is a widespread plant across southern and central Europe and adjacent parts of Asia and North Africa. It can be distinguished from all similar plants by the covering of long downy hairs that give the plant a greyish appearance. (Its close relative *Stachys byzantina* is a garden plant that sometimes escapes into the wild. *Stachys byzantina* is woollier than the downy woundwort, and even its lowest leaves are wedge-shaped, rather than heart-shaped, at the base.)

Downy woundwort has been known in Britain for centuries. The first record was from 1632, and the first herbarium specimen dates from 1733 (20 years before Linnaeus introduced the modern system of giving plants two-word scientific names).

Before the 19th century, downy woundwort occurred at scattered locations across southern and eastern England, from Hampshire and Oxfordshire to Lincolnshire and Kent. Since then, it has declined severely, and is now limited to one small area of Oxfordshire, including parts of the Wychwood Forest, an old royal hunting estate. Within this area, populations are small and scattered, but apparently stable. The area has a long history of management, and remains ecologically similar to its mediaeval state, with many of the ancient droving tracks still in use in one form or another. It seems that the enduring management of the area as a game reserve has allowed the downy woundwort to persist.

ALSO IN THE AREA

Perfoliate penny-cress, *Microthlaspi perfoliatum*, is limited in Britain to bare oolitic limestone in the Cotswolds.

CONSERVATION STATUS	VU
FLOWERING	Jul–Aug
PLANT HEIGHT	30–100 cm

Downy woundwort growing in the Vienna Woods, Austria. (CJD)

Ancient droving track in Oxfordshire. (CJD)

Water germander

Teucrium scordium

Water germander is a perennial herb of fens, dune-slacks and river banks, growing up to 50cm (20in) tall. Like the other members of the genus *Teucrium*, it has flowers with a bottom lip but no top lip, but differs in dying back in winter, having pink flowers, and having leaves that are toothed rather than cut into segments.

Water germander has a sparse distribution from westernmost Europe to Central Asia. It is fairly abundant on the shores of Lough Derg and Lough Ree in western Ireland, and formerly grew at sites over much of eastern England, but is now restricted to just three locations.

The first is at North Pit, 2.5 km north of Wicken Fen in Cambridgeshire. The population had been in long-term decline here, but a wetland creation project at nearby Kingfishers Bridge used plants propagated from North Pit, and the overall population has increased more than a hundred-fold in less than 10 years. The second site is in dune-slacks either side of the River Taw in north Devon, at Braunton Burrows and Northam Burrows. Here, too, the species was in decline until actions were taken to open up new areas of habitat by scraping away the existing tall vegetation. As a result, the population has bounced back. Conservation can, however, be an unpredictable endeavour. After water germander was discovered at the third extant location – Bassenhally Pit near Whittlesey in northern Cambridgeshire – the site was declared a nature reserve. This meant that gunmen who had previously tramped through the area in their search for wildfowl stopped doing so, and no longer opened up habitat for the water germander, which promptly died out. (It was re-introduced in 1999.) Migratory waterfowl may also be important in helping the species spread. In 2005, it was discovered at a site in Anglesey, where it was assumed to be naturally occurring, although it did not persist for long.

ALSO IN THE AREA

See under Cambridge milk-parsley (*Selinum carvifolia*, p.116) and round-headed club-rush (*Scirpoides holoschoenus*, p. 110).

CONSERVATION STATUS	EN
FLOWERING	Jul–Sep
PLANT HEIGHT	10–50 cm

Water germander growing in western Poland. (Kenraiz; WC)

Dune-slacks at Braunton Burrows. (CJD)

Twin-headed clover

Trifolium bocconei

The Lizard Peninsula in southern Cornwall is home to at least 17 species of clover, several of which are found at few other sites in Britain. The rarest of them is the twin-headed clover. It has small (4–6 mm) white flowers with no bracts, and the sepals are fused together into a tube with a total of 10 veins. Knotted clover (*Trifolium striatum*), a much more widespread coastal species, is similar to the twin-headed clover, but its sepal-tube becomes inflated when in fruit, and its leaves are hairy on their upper surfaces.

Twin-headed clover is a largely coastal species, extending along the Atlantic and Mediterranean coasts of Europe, as far as Turkey. It used to occur on Jersey, but the reduction in the rabbit population caused by an outbreak of myxomatosis reduced the amount of grazing, allowing taller plants to overgrow the clover, and it has not been seen there recently. In Britain, it is restricted to about eight localities on the Lizard Peninsula, chiefly south-facing sites over serpentine that experience droughts in summer. It is a winter annual, and flourishes best when a hot, dry summer one year is followed by a warm, wet spring the next.

Twin-headed clover was first found in Britain by the botanists Cardale Babington and William Borrer in 1839 near Cadgwith, less than 30 years after it was first described by the Italian naturalist Gaetano Savi; its scientific name commemorates the 17th-century Sicilian botanist and Cistercian monk Paolo Boccone.

ALSO IN THE AREA

Another clover found on the Lizard is the western clover, *Trifolium occidentale.* It was only in 1957 that this species was recognised as being quite different from the normal white clover, *Trifolium repens.* It is found over much of the Channel Islands, western Cornwall (including the Isles of Scilly), and sites on the coasts of north Devon and west Wales, as well as sites in Ireland and continental Europe as far south as Portugal.

CONSERVATION STATUS	VU
FLOWERING	Jun–Jul
PLANT HEIGHT	10–20 cm

Twin-headed clover growing in central Italy. (Marco Iocchi; WC)

South-facing cliffs at Cadgwith, near where twin-headed clover was first found in Britain. (Bill Boaden; GG)

Upright clover

Trifolium strictum

Upright clover is, despite its name, quite a short plant, normally growing only a few inches high in Britain. Superficially, upright clover looks quite similar to a number of other low-growing clovers. Its stipules (membranous to leaf-like outgrowths from the point where the base of a leaf meets the stem) are distinctively saw-toothed at the margins, and both they and the leaves have rounded glands at the tips of their teeth.

Upright clover is a winter-annual Atlantic–Mediterranean species, with a range that extends from Morocco to Turkey, and northwards into various parts of Europe. It is one of several rare clovers that occur on the Lizard Peninsula (see twin-headed clover, *Trifolium bocconei*, p. 128), where it was first discovered by the Rev. C. A. Johns. Johns marvelled at the diversity of legumes at Caerthillian Cove, where he was able to find six clover species and two other legumes in an area small enough to cover with his hat. As he reported in *The Phytologist*, 'had the rim been a little wider, I might have added *Genista tinctoria* and *Lotus corniculatus*'. (Oddly, in his *A Week at the Lizard*, published the following year, he claimed only to have covered half as many species with his impromptu quadrat.)

The Cornish populations are almost all in protected sites and are grazed to maintain the short sward that upright clover requires. A small and unstable population occurs at Stanner Rocks in Radnorshire. It varies in number from year to year, sometimes dropping to zero before reappearing from the seed bank. Recent annual estimates have produced counts such as 6 and 13.

A further record was made in 1849 based on a herbarium specimen collected 12 years earlier on Anglesey, where it grew 'in abundance, covering a space of fifty yards square'. The species has not been seen there since, and there must be some suspicion that the plant was misidentified. There is also a population on Jersey, but the species seems to have died out on Guernsey.

CONSERVATION STATUS VU
FLOWERING Jun–Jul
PLANT HEIGHT 10–25 cm

ALSO IN THE AREA

See under fringed rupturewort (*Herniaria ciliolata*, p. 64) for the Lizard, and Radnor lily (*Gagea bohemica*, p. 56) for Stanner Rocks.

Upright clover growing in Brittany, France. (Geneviève Botti; TB)

Coast near Caerthillian Cove, Cornwall. (CJD)

Spring speedwell

Veronica verna

Breckland, or 'the Brecks', is a special area on the Norfolk–Suffolk border that is home to the country's only inland sand dunes. The climate there is very dry for Britain and very continental, which allows plants that are more commonly found on steppes to survive here.

Spring speedwell is one of three species of tiny speedwell restricted to Breckland, and the only one that is native. Both the Breckland speedwell, *Veronica praecox*, and the fingered speedwell, *Veronica triphyllos*, were introduced from continental Europe several hundred years ago.

Spring speedwell was found by the local aristocrat Sir John Cullum in 1775, and is very easily overlooked; its flowers are less than 3 mm in diameter and fall from the plant at the slightest touch, and the whole plant may be only 3 cm (1 in) tall, or up to 15 cm (6 in). Its leaves have deep lobes on either side, something it shares with the fingered speedwell, but the two can be told apart by the fruit stalk, which is much shorter than the adjacent bract in the spring speedwell, but much longer in the fingered speedwell.

Even within Breckland, the distribution of the spring speedwell is restricted to about a dozen sites near the village of Icklingham. Each of the populations is small, ranging from a single individual to around 50. Given that the species is strictly an annual, this makes it very likely that it will be lost from some of these sites in the coming years.

ALSO IN THE AREA
See under Suffolk lungwort, *Pulmonaria obscura* (p. 96).

CONSERVATION STATUS	EN
FLOWERING	Apr–Jun
PLANT HEIGHT	3–15 cm

Spring speedwell. (Julia Kruse; IC)

Lakenheath Warren. (CJD)

Dwarf pansy

Viola kitaibeliana

Dwarf pansy is the smallest of the native British pansies, with flowers only 4–8 mm (0.15–0.30 in) high. As such, it is unlikely to be confused with anything other than aberrant forms of the field pansy, *Viola arvensis*, or the wild pansy, *Viola tricolor*. It is found across central and eastern Europe, from Switzerland, Austria and Slovakia southwards and eastwards to the Middle East and the Black Sea, as well as along the Atlantic coast from Portugal to near Calais in France (and perhaps into Belgium), including the Channel Islands. A very similar pansy, once considered a variety of the dwarf pansy, occurs across most of North America; it is now generally considered to be a separate species, *Viola bicolor*, known locally as the 'field pansy' (the only North American native pansy).

In Britain, the dwarf pansy is only found on the Isles of Scilly, and within that archipelago, only on the islands of Bryher and Teän, with a single recent record from the island of St. Martin's. It grows in relatively open, usually sandy grassland, and generally where rabbits closely crop the vegetation. It flourishes especially on the edges of paths, where the trampling reduces competition and so favours the dwarf pansy. It was first found in 1873, but not correctly identified until 1901.

One of the best recent years for the dwarf pansy followed the 'Great Storm' of 1987, which moved a lot of sand across the islands. Although it benefits from the exceptionally mild climate of the Scilly Isles, the dwarf pansy does not tolerate salt, and hardly ever grows within 30 m (100 ft) of the high-water mark.

ALSO IN THE AREA

See under least adder's-tongue, *Ophioglossum lusitanicum* (p. 78).

CONSERVATION STATUS	NT
FLOWERING	Apr–Jul
PLANT HEIGHT	2–10 cm

Dwarf pansy growing in southern France. (Bernard Dupont; WC)

Coastal grassland on Bryher, Isles of Scilly. (Andrew Abbott; GG)

Fen violet

Viola persicifolia

Fen violet (*Viola persicifolia*, but sometimes still referred to by its old name, *Viola stagnina*) is a species of violet found across the more central parts of Europe. It is rare over much of its distribution, probably due to the changing face of agriculture.

Fen violet has roots that creep underground and, at intervals, send up shoots without rosettes of leaves at their bases. Its flowers are 10–15 mm (0.4–0.6 in) wide and very pale blue, with a green-tinged spur pointing backwards, and its leaves are considerably longer than wide. (The scientific name *persicifolia* reflects the supposed similarity of the leaves to those of peach trees.)

Fen violet has been recorded in the past from damp calcareous grassland from Cambridgeshire to the Humber, but has died out in most places, and is now known at only three sites – Wicken Fen and Woodwalton Fen in Cambridgeshire, and Otmoor in Oxfordshire. In Ireland, fen violet grows in turloughs (temporary lakes over limestone) around Lough Erne and in the Burren. The seeds are long-lived, allowing the plant to reappear following disturbance. For example, it was not seen for 10 years at Wicken Fen before it reappeared there in 2014. The Otmoor population was also thought to have died out, having been apparently absent for more than 30 years before a small population was rediscovered in 1997 after the clearance of some willow scrub. This provides some hope for the species' survival in Britain, given that its numbers at Woodwalton Fen have dipped as low as one.

ALSO IN THE AREA

See under Cambridge milk-parsley, *Selinum carvifolia* (p. 116).

CONSERVATION STATUS	EN
FLOWERING	Apr–Jun
PLANT HEIGHT	10–25 cm

Fen violet growing in Alsace, France. (Hugues Tinguy; TB)

Woodwalton Fen. (Hugh Venables; GG)

Critical groups

Lady's-mantles

Alchemilla spp.

Lady's-mantles are common as garden plants, especially *Alchemilla mollis*, soft lady's-mantle. They have dense groups of rather odd flowers with four greenish sepals and no petals, and the various species are usually told apart by fine details of the lobed leaves.

Twelve of the 15 microspecies present in Britain are native. *Alchemilla micans* is only found in Britain near Hadrian's Wall in Northumberland, and *A. subcrenata* is limited to the moors of County Durham, although both species also occur in other countries. *Alchemilla minima*, on the other hand, is a British endemic, growing only on the slopes of Ingleborough and Whernside in the Yorkshire Dales.

Helleborines

Epipactis spp.

Although not strictly apomictic, one group of orchids has abandoned sexual reproduction in favour of self-fertilisation. This has resulted in several reproductively isolated lineages, often restricted to small areas, difficult to tell apart, and of unclear taxonomic rank. They include the fairly widespread narrow-lipped helleborine of southern England (*Epipactis leptochila*), as well as rarer taxa such as the dune helleborine (*Epipactis dunensis*) of North Wales and north-western England, the Tyne helleborine (*Epipactis youngiana*) either side of the English–Scottish border, and the Lindisfarne helleborine (*Epipactis sancta*), which is only found on the tidal island of Lindisfarne off the Northumbrian coast.

Eyebrights

Euphrasia spp.

Eyebrights are small, hemiparasitic plants, now placed in the broomrape family, along with all the other parasitic and hemiparasitic plants that were formerly placed in the figwort family. Eyebrights parasitise a wide range of other plants, and live mainly in open grassland.

There are 18 native species of eyebright in Britain, of which seven are found nowhere else, and dozens of hybrids between them. The rarest species is probably Pugsley's eyebright, *Euphrasia rotundifolia*, which is only found in its pure form at Portskerra on the north coast of mainland Scotland, where it grows alongside the Scottish primrose, *Primula scotica*.

Hawkweeds

Hieracium spp.

Hawkweeds are yellow-flowering members of the daisy family, usually distinguishable from hawk's-beards (*Crepis* spp.) by their general hairiness and the gradual, not abrupt, change from the outermost phyllaries (bracts around the flower-head) to the innermost. More than 400 microspecies have been recorded from the British Isles, a large number of which are endemic, and many of which have very restricted distributions.

Sea-lavenders

Limonium spp.

Sea-lavenders are perennial herbs in the same family as thrift (*Armeria maritima*) that grow in coastal habitats in the more southerly parts of Britain. Their branching stems bear ranks of blue-purple flowers that are quite attractive when seen up close, but are easily missed against the mud. The genus has been subject to some taxonomic confusion, with the number of described taxa (species and subspecies) increasing massively in the late 20th century. As a result, the distributions of these poorly distinguishable taxa are not well known, although many of them appear to be rare. *Limonium loganicum* is only found on the Penwith Peninsula, Cornwall, and *L. paradoxum*, *L. parvum* and *L. transwallianum* are each found on one peninsula in Pembrokeshire. Many of the subspecies of other sea-lavender species are similarly restricted.

Brambles

Rubus spp.

Brambles are familiar and widespread plants, with their prickly, trailing stems, (usually) five-parted leaves and tasty fruit. They need pollen to be deposited by a pollinator in order for their fruit to set, although in most cases no DNA from the pollen is used. There are more than 300 microspecies in the British Isles including around 200 that are endemic to these islands, and many of which are restricted to very small areas.

Apart from brambles, several other species of *Rubus* grow in Britain, including the cloudberry (*Rubus chamaemorus*) and raspberry (*Rubus idaeus*).

Whitebeams

Sorbus spp.

Whitebeams are small trees with white flowers and red fruits, closely related to the rowan or mountain ash, *Sorbus aucuparia*. They are associated with calcareous rocks, and some species are very widespread while others are very restricted in their distribution.

In Scotland, the Arran whitebeams, *Sorbus arranensis*, *S. pseudofennica* and *S. pseudomeinichii*, are all limited to the Isle of Arran.

In Wales, *S. minima*, *S. leyana*, *S. leptophylla* and *S. cambrensis* all grow at single sites in southern Brecknockshire; *S. stenophylla* grows in the Llanthony Valley; *S. stirtoniana* grows in Montgomeryshire; *S. cuneifolia* grows at a site near Llangollen.

In England, Cheddar Gorge is the only place where *S. eminentoides*, *S. cheddarensis* and *S. rupicoloides* grow; the Avon Gorge is the only place for *S. bristoliensis*, *S. leighensis* and *S. wilmottiana*; *S. margaretae*, *S. vexans* and *S. subcuneata* are all limited to the Exmoor coast; *S. admonitor* was first discovered near Lynmouth with a 'No Parking' sign nailed onto it (hence the name *admonitor*), and is only found in that area.

Dandelions

Taraxacum spp.

Dandelions are familiar garden weeds, with their bold, yellow flower-heads, coarsely toothed leaves and 'clocks' of wind-borne seeds.

There are around 230 dandelion microspecies in the British Isles, including at least 35 that are found nowhere else in the world, growing in all sorts of habitats from dune systems to high mountains.

Taraxacum breconense is endemic to the Brecon Beacons, *T. hirsutissimum* to the southernmost area of Shetland's Mainland, *T. serpenticola* to Unst, *T. pseudonordstedtii* to the North Pennines, and *T. margettsii* to the Lizard Peninsula, and many other species also have very narrow distributions in Britain.

Image sources:

Arabis alpina	http://www.geograph.org.uk/photo/2005037
Arabis scabra	http://www.tela-botanica.org/bdtfx-nn-5973-illustrations
Atriplex pedunculata	http://www.tela-botanica.org/bdtfx-nn-8435-illustrations
Carex flava	http://dbiodbs.units.it/carso/chiavi_pub26?spez=934
Carex microglochin	http://dbiodbs.units.it/carso/chiavi_pub26?spez=4357
Carex microglochin	http://www.geograph.org.uk/photo/3693526
Cephalanthera rubra	http://www.tela-botanica.org/bdtfx-nn-15696-illustrations
Corrigiola litoralis	https://commons.wikimedia.org/wiki/File:CorrigiolaLitoralis2.jpg
Crassula aquatica	http://www.geograph.org.uk/photo/1998299
Crepis praemorsa	https://commons.wikimedia.org/wiki/File:Crepis_praemorsa_(Trauben-Pippau)_IMG_0429.JPG
Damasonium alisma	http://www.tela-botanica.org/bdtfx-nn-21525-illustrations
Dianthus gratianopolitanus	http://dbiodbs.units.it/carso/chiavi_pub26?spez=12325
Dianthus gratianopolitanus	http://www.geograph.org.uk/photo/2767287
Diapensia lapponica	https://commons.wikimedia.org/wiki/File:Diapensia_lapponica_upernavik_2007-07-11_2.jpg
Draba aizoides	http://www.tela-botanica.org/bdtfx-nn-75099-illustrations
Epipogium aphyllum	http://www.geograph.org.uk/photo/3290308
Epipogium aphyllum	https://commons.wikimedia.org/wiki/File:Epipogium_aphyllum-002.jpg
Gagea bohemica	https://commons.wikimedia.org/wiki/File:Gagea_bohemica_sl8.jpg
Gentiana nivalis	http://www.geograph.org.uk/photo/2355866
Gentiana nivalis	https://commons.wikimedia.org/wiki/File:Gentiana_nivalis_(habitus).jpg
Hydrilla verticillata	https://www.flickr.com/photos/bigcypressnps/30908170624
Isoetes histrix	http://dbiodbs.units.it/carso/chiavi_pub26?spez=6469
Juncus pygmaeus	http://www.tela-botanica.org/bdtfx-nn-36655-illustrations
Ophioglossum lusitanicum	http://dbiodbs.units.it/carso/chiavi_pub26?spez=8085
Ophioglossum lusitanicum	http://www.geograph.org.uk/photo/3092593

Orobanche caryophyllacea	https://commons.wikimedia.org/wiki/File:Orobanche_caryophyllacea_080608.jpg
Oxytropis campestris	https://commons.wikimedia.org/wiki/File:Oxytropis_campestris01.jpg
Petrorhagia nanteuilii	http://www.tela-botanica.org/bdtfx-nn-48404-illustrations
Phyllodoce caerulea	https://commons.wikimedia.org/wiki/File:Phyllodoce_caerulea_Kilpisjärvi_2012-12a.jpg
Phyllodoce caerulea	https://commons.wikimedia.org/wiki/File:The_Sow_of_Atholl_from_the_A9_road.jpg
Potamogeton epihydrus	http://potamogetonaceae.e-monocot.org/gallery
Potamogeton epihydrus	http://www.geograph.org.uk/photo/2121959
Romulea columnae	http://www.geograph.org.uk/photo/5099833
Romulea columnae	http://www.tela-botanica.org/bdtfx-nn-75701-illustrations
Saxifraga cernua	http://www.geograph.org.uk/photo/3571393
Saxifraga cernua	https://commons.wikimedia.org/wiki/File:Saxifraga_cernua_plant_upernavik_2007-07-09.jpg
Schonoplectus triqueter	https://commons.wikimedia.org/wiki/File:Schoenoplectus_triqueter_kz1.JPG
Schoenus ferrugineus	http://www.tela-botanica.org/bdtfx-nn-61423-illustrations
Schoenus ferrugineus	https://commons.wikimedia.org/wiki/File:Ben_Vrackie_with_the_lake.JPG
Scleranthus perennis	http://www.geograph.org.uk/photo/2541361
Selinum carvifolia	http://dbiodbs.units.it/carso/chiavi_pub26?spez=2730
Senecio paludosus	https://commons.wikimedia.org/wiki/File:Senecio_paludosus_kz1.jpg
Silene suecica	https://commons.wikimedia.org/wiki/File:Viscaria_alpina002.jpg
Teucrium scordium	https://commons.wikimedia.org/wiki/File:Teucrium_scordium_kz1.jpg
Trifolium bocconei	http://www.geograph.org.uk/photo/4143524
Trifolium bocconei	https://commons.wikimedia.org/wiki/File:Trif_bocconei33.JPG
Trifolium strictum	http://www.tela-botanica.org/bdtfx-nn-69436-illustrations
Veronica verna	http://dbiodbs.units.it/carso/chiavi_pub26?spez=6236
Viola kitaibeliana	http://www.geograph.org.uk/photo/3087844
Viola kitaibeliana	https://commons.wikimedia.org/wiki/File:Dwarf_Pansy_%28Viola_kitaibeliana%29_%288524983666%29.jpg
Viola persicifolia	http://www.geograph.org.uk/photo/3384760
Viola persicifolia	http://www.tela-botanica.org/bdtfx-nn-74554-illustrations

Index